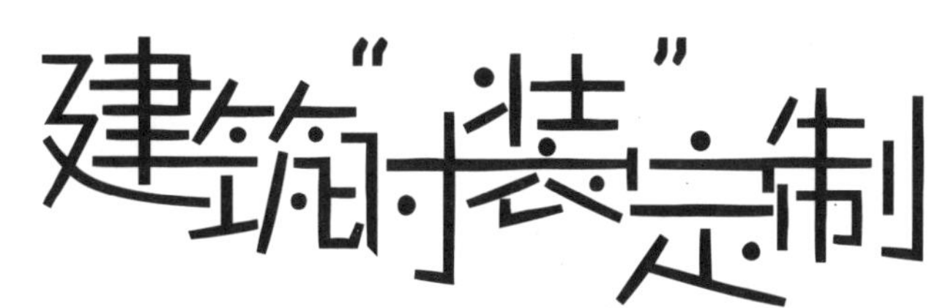

FASHION IN ARCHITECTURE

混凝土

CONCRETE

凤凰空间·北京 编

江苏科学技术出版社

谨以此书高调迎接建筑立面的个性化时代，或者说建筑
“时装定制”时代的到来。

This book is dedicated to the forthcoming era of individualized building facades, or the era of "custom fashion" for architecture.

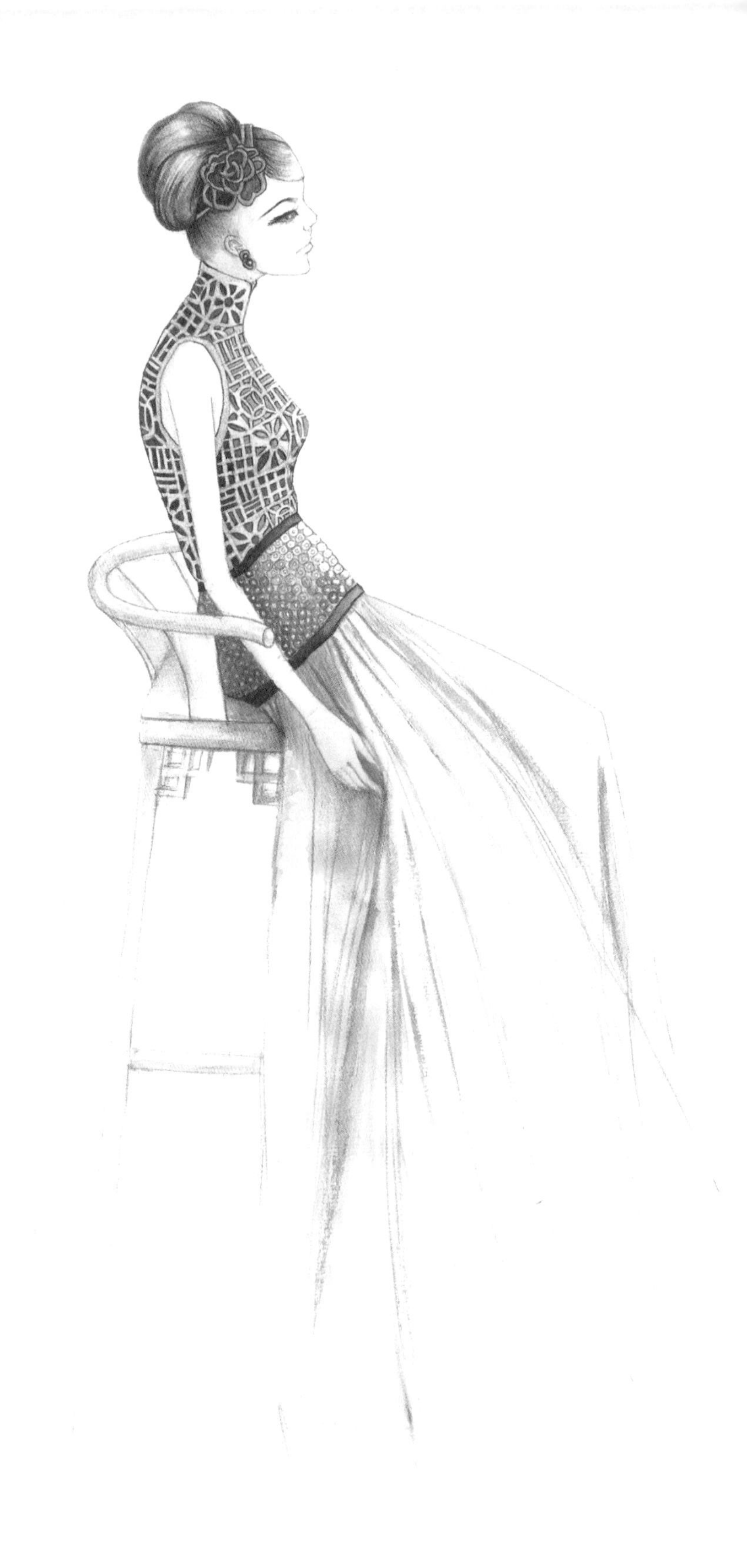

前　言
PREFACE

■ 关于混凝土，可追溯到古代。古罗马人将火山灰混合石灰、砂制成的天然混凝土用于建筑中。

■ 1824 年，英国建筑工人约瑟夫 · 阿斯普丁发明了波特兰水泥并获得专利。

■ 法国园艺家约瑟夫 · 莫尼埃发明了钢筋混凝土，并于 1867 年获得专利。1875—1877 年，他主持建造了第一座人行钢筋混凝土桥。

■ 1900 年，万国博览会上展示了钢筋混凝土在很多方面的应用，在建材领域引起了一场革命。

■ 1918 年，达夫 · 艾布拉姆斯发表了著名的计算混凝土强度的水灰比理论。钢筋混凝土开始成为改变这个世界景观的重要材料。

■ 如今，材料、技术、设备、工艺不断地推陈出新，混凝土作为建筑中的无冕之王，在建筑立面中的运用也不甘落后。近年来，世界各地涌现出精彩纷呈的独具混凝土魅力的特色建筑。

■ The history of concrete can be traced back to ancient times. The ancient Romans used natural concrete combined of lime, pozzolan ash, sand and gravel in buildings.

■ In 1824, a British stone mason, Joseph Aspdin invented Portland cement and obtained a patent.

■ Reinforced concrete was invented by a French gardener, Joseph Monier, who received a patent in 1867.

■ At the World Exposition of 1900, the extensive utilization of concrete brought an innovation in the field of construction material

■ In 1918, Duff Abrams published a well-known computational theory of water-cement ratio to determine the concrete strength. Reinforced concrete became an important material to change the world landscape.

■ With the development of new material, technique, equipment and technology, concrete has become a universal dominant construction material, and also applied in the façade of architecture. In recent years, a huge number of significant concrete architecture were created worldwide.

目 录
CONTENTS

第一部分 PART 1

建筑混凝土“时装”历史

HISTORY OF CONCRETE "FASHION" IN ARCHITECTURE

在这里，我们一起来回顾一下建筑中混凝土立面设计的历史，带领大家穿越时空，共同领略建筑“时装”的美，领略建筑“时装”设计师们的别具匠心，体味带有混凝土质感的建筑奇观。

Here, let's look back over the history of concrete façade in architecture, appreciate the magnificent concrete "fashion" and the creativeness of architects, and enjoy the concrete architecture wonders across time and space.

1. 古代

混凝土在世界范围内被广泛地应用。中国甘肃省大地湾发现了古代混凝土的遗迹。尽管这所住宅遗迹距今已有 5000 年，但人们却精确地发现在地板中使用了类混凝土的材料。这里要先明确一下混凝土的定义，一般来说，由水、水泥、细骨料（砂）、粗骨料（砂石）混合而成的建筑材料被称为混凝土。在遗址中，也发现了可以烧制水泥的窑。这些水泥可能是由以碳酸钙与黏土为主要成分的料姜石制成的。

正式将混凝土作为建筑材料的是古代罗马人。他们用火山灰混合石灰、砂制成的天然混凝土用于建筑中，建造了街道、桥梁、神殿、礼堂、广场等各种各样的建筑物。据说直到罗马帝国为止，这些社会基础设施都在持续地受到维护。正是因为古代混凝土能让建筑物更加坚固、雄伟，所以这样的成就在今天看来也是令人钦佩的。也正因为这些建筑，才支撑起了强大的古代罗马文明。即使到了今天，我们仍然还在使用“罗马非一朝一夕建成”和“条条大路通罗马”这样的谚语。

大地湾古代混凝土遗址，中国，甘肃　Ancient Concerete Site at Dadi Bay, Gansu, China

1. ANCIENT

Concrete has found very wide application worldwide in construction field. Ancient concrete site was found in Dadi Ba in Gansu, China. Though this Dadi Bay Relic site has a history of 5000 years, it is found that concrete was used for floors. Here, we first give a clear definition of concrete. Generally speaking, concrete is a composite construction material made primarily with water, cement, sand and gravel. A kiln used to fire cement was also found in the Dadi Bay Relic site. The concrete is probably made from ginger-like rock whose main components are calcium carbonate and clay.

Concrete was formally used as construction material in ancient Rome. The ancient Romans used natural concrete combined of lime, pozzolan ash, sand and gravel in various constructions, such as streets, bridges, temple, auditorium, forum, etc. It's said that these infrastructures were still under maintenance until the Roman Empire. Concrete made the constructions more stately and grand, and these achievements are still impressive today. It's these constructions that support the ancient Roman civilization. Until today, we still use the proverbs, such as "Rome wasn't built in a day" and "All roads lead to Rome".

2. 近代

1824 年，英国建筑工人约瑟夫 · 阿斯普丁发明了波特兰水泥并获得专利。由于用它配制成的混凝土具有工程所需要的强度和耐久性，而且操作简单、原料易得、造价较低，特别是能耗较低，因而用途极为广泛。

钢筋混凝土是现代常用的建筑结构材料，由法国园艺家约瑟夫 · 莫尼埃发明，并于 1867 年获得专利。1875 年，莫尼埃主持建造了世界上首座钢筋混凝土大桥。法国工程师埃纳比克在 1867 年的巴黎博览会上看到莫尼埃用铁丝网和混凝土制作的花盆、浴盆和水箱后受到启发，于是设法把这种材料应用于房屋建筑上。1879 年，埃纳比克开始制造钢筋混凝土楼板，开发了一套由钢筋箍和纵向杆加固的混凝土结构梁构成的建筑体系，并于 1892 年获得专利。1884 年，德国建筑公司购买了莫尼埃的专利，进行了第一批钢筋混凝土的科学实验，研究了钢筋混凝土的强度、耐火能力以及钢筋与混凝土的黏结力。

1887 年，德国工程师科伦首先发表了钢筋混凝土的计算方法；英国人威尔森申请了钢筋混凝土板专利；美国人海厄特对混凝土横梁进行了实验。1900 年，万国博览会上展示了钢筋混凝土在很多方面的应用，特别是在建材领域引起了一场革命。

约瑟夫·莫尼埃主持建造的首座钢筋混凝土桥，法国
The First Reinforced Concrete Bridge Designed by Joseph Monier, France

2. MODERN

In 1824, a British stone mason, Joseph Aspdin invented Portland cement and obtained a patent. Since the concrete mixed of Portland cement has the properties of high strength and durability which are needed in construction. And it has the advantages of simple operation, easy obtainment of the raw material, low cost and low energy consumption, so it has wild application.

Reinforcement concrete is wildly used architectural constructional material. It was invented by a French gardener, Joseph Monier, who received a patent in 1867. In 1875, Monier designed the first reinforced concrete bridge at the Castle of Chazelett, France. François Hennébique, a French engineer, saw Monier's reinforced concrete flowerpots, tubs and tanks at the 1867 Paris Exposition, and began experimenting with ways to apply this new material to building construction. He began with reinforced-concrete floor slabs in 1879, and later developed as a set of building the structure of reinforced concrete beams reinforced hoop and longitudinal rod, patented in 1892, using structural beams of concrete reinforced with stirrups and longitudinal bars designed to resist the tensile forces. In 1884, German construction company bought Monier's patent, and conducted the first scientific experiment of reinforced concrete, studying the strength and the fire resistance of reinforced concrete, and its bond strength between bars and concrete.

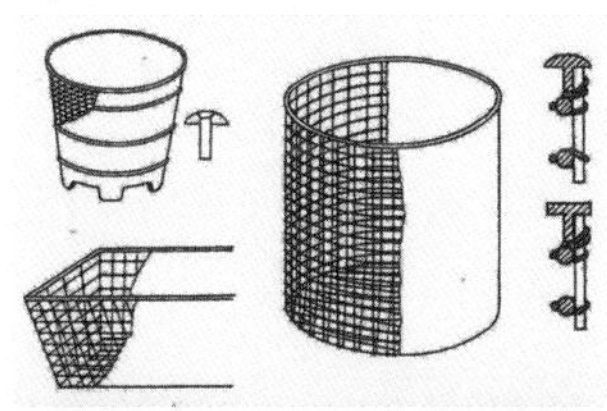

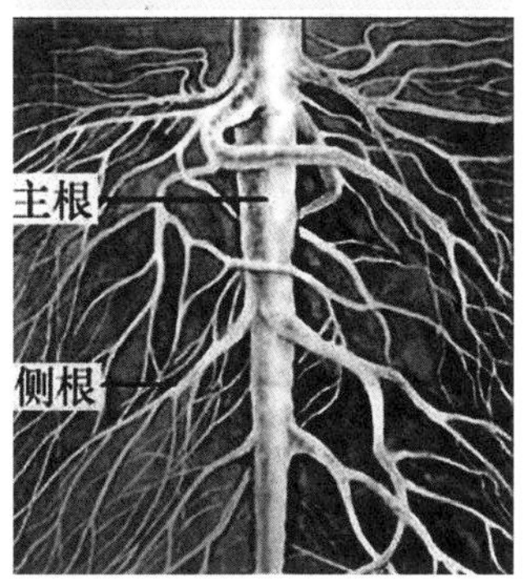

约瑟夫·莫尼埃发明了“钢筋混凝土”，其灵感来自植物的根系。
Inspired by plant roots, Joseph Monier invented the reinforced concrete.

In 1887, German engineer Cologne first published the method of calculation of the reinforced concrete; Englishman Wilson applied for a reinforced concrete slab patent; an American, Hyatt carried out experiments on beams. At the World Exposition of 1900, the extensive utilization of concrete brought an innovation in the field of construction material. At World's Fair of 1900, the extensive utilization of concrete brought an innovation in construction material field.

3. 现代

1918 年，达夫 · 艾布拉姆斯发表了著名的计算混凝土强度的水灰比理论。于是，钢筋混凝土开始成为改变这个世界景观的重要材料。此后，便相继出现了轻集料混凝土、加气混凝土以及其他类型的混凝土，各种混凝土外加剂也开始被广泛使用。

自 20 世纪 60 年代以来，混凝土领域中有了高效减水剂，随之相应的出现了流态混凝土这一种类。同时，高分子材料也进入到混凝土材料领域中来，出现了聚合物混凝土。此外，多种纤维被用于分散配筋的纤维混凝土也相继出现。如今，现代测试技术也越来越多地应用于混凝土材料科学的研究当中。材料、技术、设备以及工艺都在不断地推陈出新，使得混凝土早已成为建材中的无冕之王。

本书将注意力更多地凝聚在混凝土材料在建筑立面上的应用。如何利用混凝土本身的可塑性赋予建筑崭新的风貌，属于"新混凝土"的应用范畴。本书收集了国际上以混凝土为建筑材料的新设计、新运用的建筑案例，为读者奉上一场靓丽非凡的"混凝土建筑时装秀"。

两百年纪念市政中心，阿根廷，科尔多瓦
Bicentennial Civic Center, Córdoba, Argentina

3. CONTEMPORARY

In 1918, Duff Abrams published a well-known computational theory of water-cement ratio to determine the concrete strength. Reinforced concrete became an important material to change the world landscape. Later, lightweight aggregate concrete, aerated concrete and other types of concrete, as well as a variety of concrete admixtures have begun to be wildly used.

Since the 1960s, the water-reducing agent has been used at mixing for casting concrete, and the flowing concrete has been created. Then, polymer material has appeared in the field of concrete materials, and polymer concrete has been created. Besides, fiber reinforced concrete which uses a variety of fibers for dispersion of reinforcement. Today, modern testing techniques are also increasingly used in the study of concrete materials. With the development of new material, technique, equipment and technology, concrete has become a universal dominant construction material.

This book put more focus on the application of concrete in building façade. It is the application of new concrete to give the building a new look by using the flexibility of concrete. In this book, we collected the latest architecture projects in the world with new application and design of concrete, presenting the readers with a gorgeous fashion show of concrete architecture.

4．未来

随着现代材料科学的不断进步，作为最主要的建筑材料之一的混凝土已逐渐向高强度、高性能、多功能和智能化发展，用它建造的混凝土结构也趋于大型化和复杂化。绿色高性能混凝土是未来混凝土的发展趋势，可实现高效可循环利用。

混凝土已经融入我们的生活，无时无刻不在影响着我们的今天。从混凝土的发展状况可以看出，混凝土是一种充满生命力的建筑材料，正在以令人瞠目的速度发展着，为未来的建筑创造无限的空间和可能。

科幻中的未来月球城市 Moon City in Science Fiction

4. FUTURE

With the continuous development of modern materials science, concrete, as one of the most important construction materials, is gradually developing towards high-strength, high-performance, multi-function and intellectualized feature. Concrete structure also tends to be large-scale and complicated. In the future, high-performance green concrete will become the future trend of concrete, and the construction should make effective and circular use of it.

Concrete plays a vital part in our daily lives, and it affects us in every corner. From its development status, we find that concrete is a construction material full of vitality. And it is developing in a quick speed, creating infinite space and possibility for the future architecture.

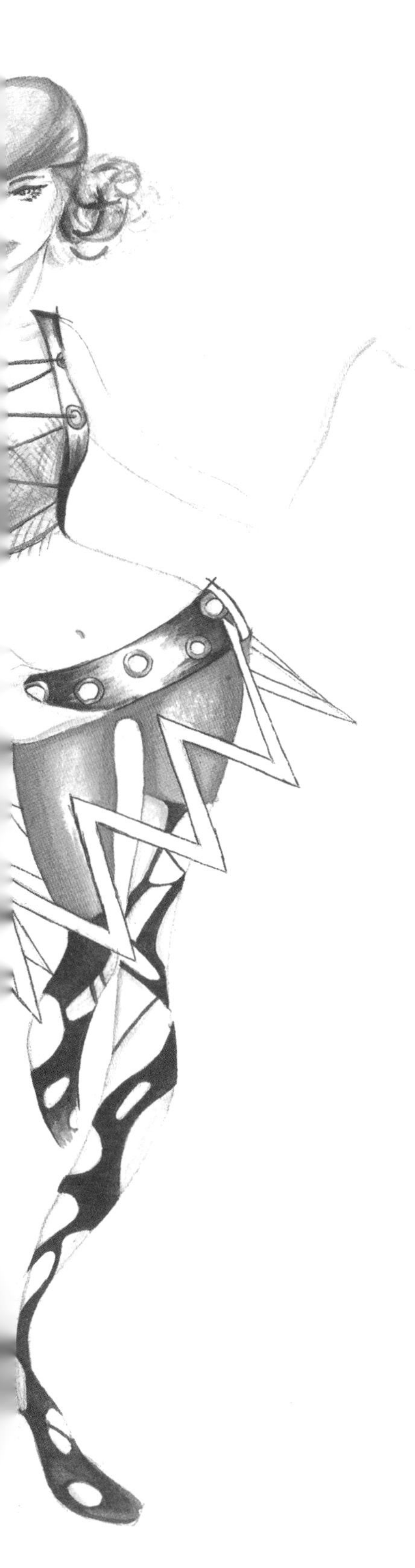

第二部分 PART 2

建筑混凝土“时装”案例

CASES OF CONCRETE "FASHION" IN ARCHITECTURE

在这里，我们将注意力更多地凝聚在混凝土材料在建筑立面上的应用。如何利用混凝土本身的可塑性赋予建筑崭新的风貌，属于“新混凝土”的应用范畴。本书收集了国际上以混凝土为建筑材料的新设计、新运用的建筑案例，为读者奉上一场靓丽非凡的“混凝土建筑时装秀”。

Here, we put more focus on the application of concrete in building façade. It is the application of new concrete to give the building a new look by using the flexibility of concrete. We collected the latest architecture projects in the world with new application and design of concrete, presenting the readers with a gorgeous fashion show of concrete architecture.

利物浦百货商场

LIVERPOOL VILLAHERMOSA

项目名称：
利物浦百货商场
Project Name:
LIVERPOOL VILLAHERMOSA

项目设计： Iñaki Echeverria
项目地点： 墨西哥，塔巴斯科
立面材料： 混凝土
竣工时间： 2011 年
摄影： Luis Gordoa

Architects: Iñaki Echeverria
Location: Tabasco, Mexico
Façade Material: concrete
Completion: 2011
Photographs: Luis Gordoa

线条的变化使服装的立体感突出，更具“建筑美感”。这件设计奇巧的礼裙，仿佛由若干条带拼接而成，如百叶那般开合。条带扭曲形成鼓起的褶皱和镂空，造就时装的立体感。镂空处若隐若现显露出里衬，黑白的色彩搭配增加层次及变化；腰间点缀的金色腰带，更勾画出曼妙的腰线，妩媚动人。利物浦百货商场的新立面由螺旋桨形的混凝土预制件构筑而成，每一片螺旋桨均绕轴旋转，动感十足。

The variations in lines highlight the dress's three-dimensional effect. This exquisite dress is seemingly composed of a number of stripes which are open or close like the shutter. The pleats and hollow-out created by twisted stripes bring a three-dimensional effect to this fashionable dress. The lining appears indistinctly through the hallow-out and the black-white pairing adds more layering and variations to the dress. With a golden belt embellishing the waist, it looks more graceful. The new façade of Liverpool department store is built by the precast concrete pieces shaped like a propeller. Each propeller rotates on its axis, creating a sense of movement.

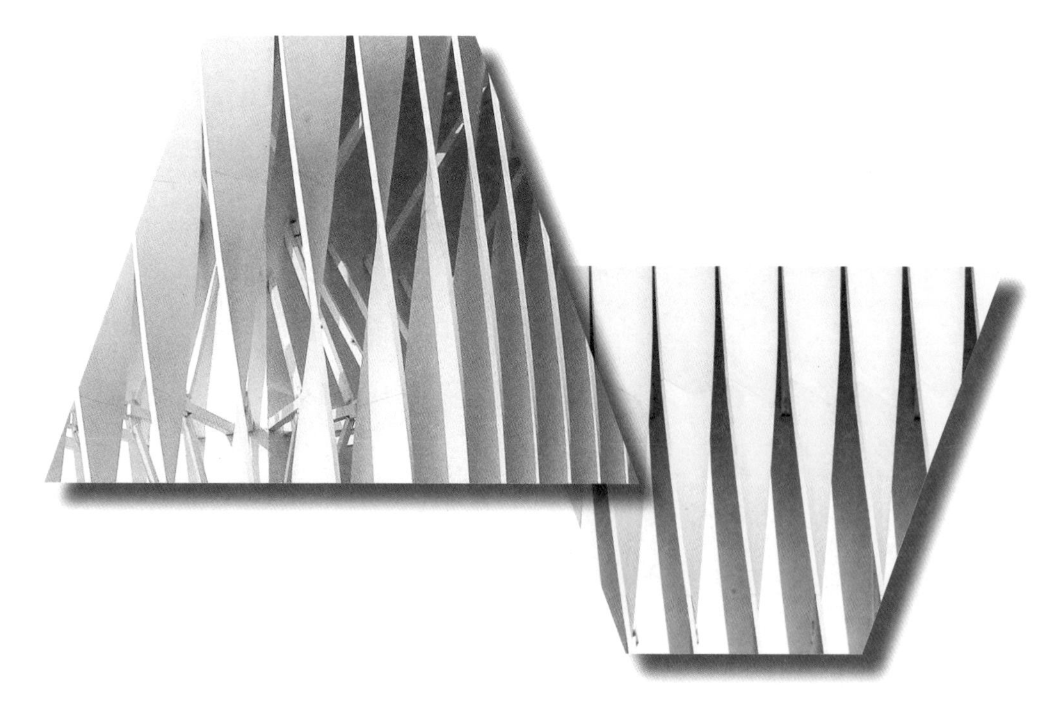

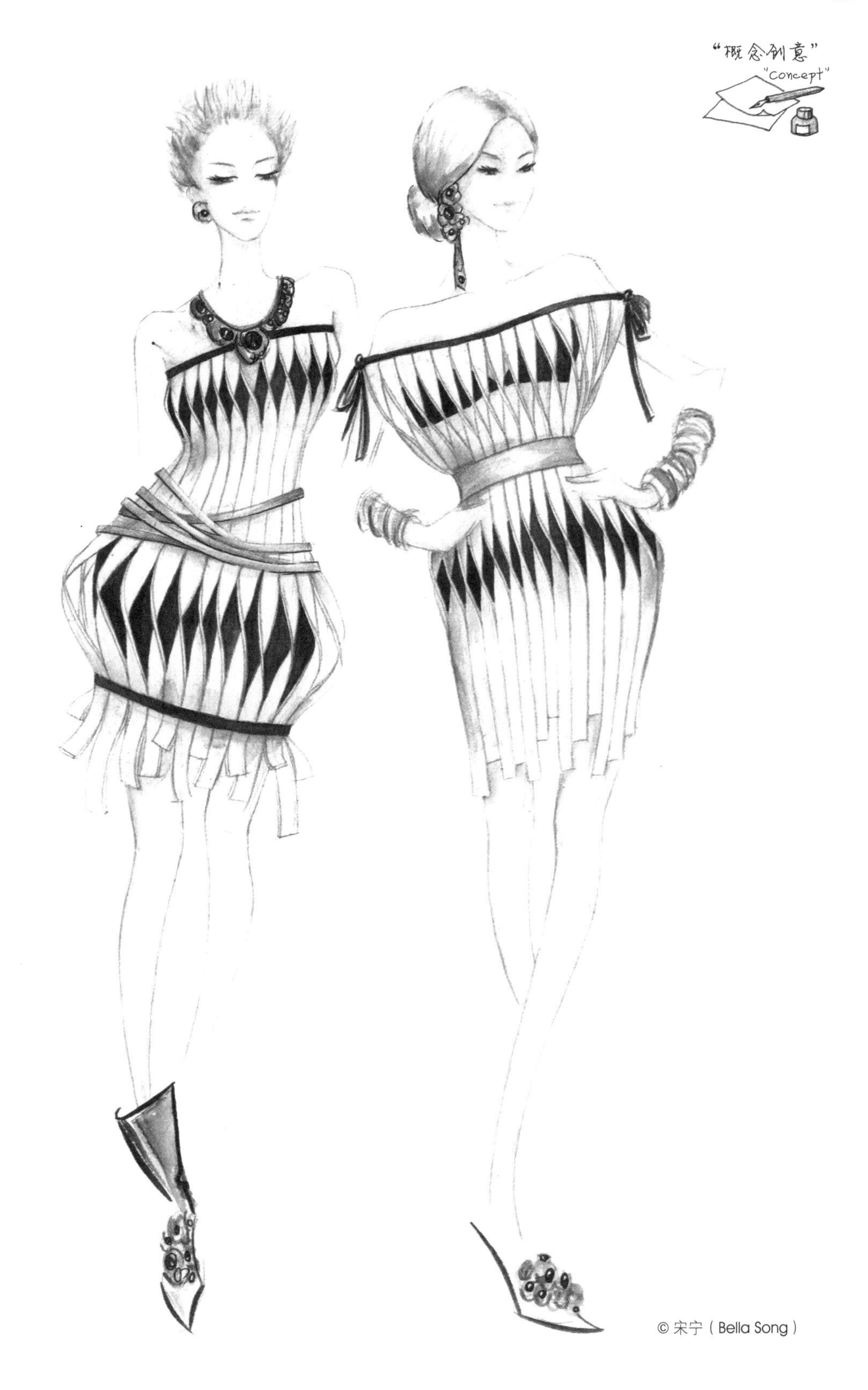

© 宋宁（Bella Song）

"面料定制"
"Material"

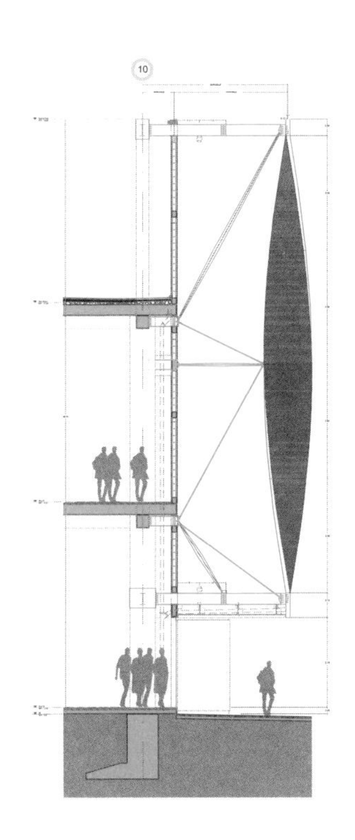

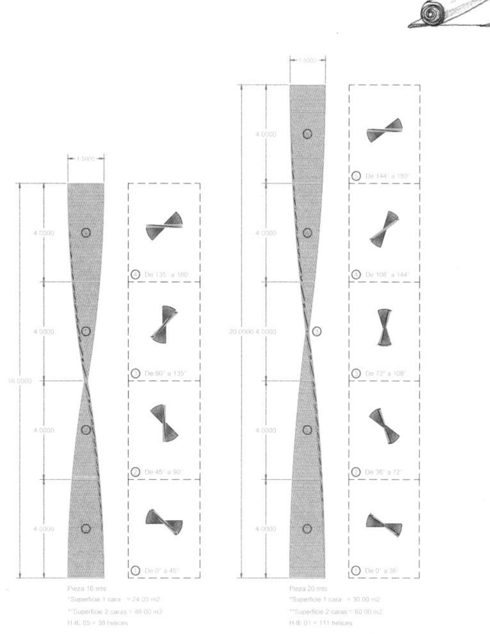

立面详图 FAÇADE DETAIL

鉴于当地的热带气候，太阳光强烈且湿度也高。因此，混凝土被选做外立面的主材，耐用且能够经得起时间的考验。建筑师采用了创新型建筑技术，力求使利物浦百货商场呈现全新的形象。立面由五种螺旋桨形的预制件构筑而成，每一片螺旋桨均绕轴旋转180°。这些桨片长度为16~20 m不等，其长度取决于所处的位置。

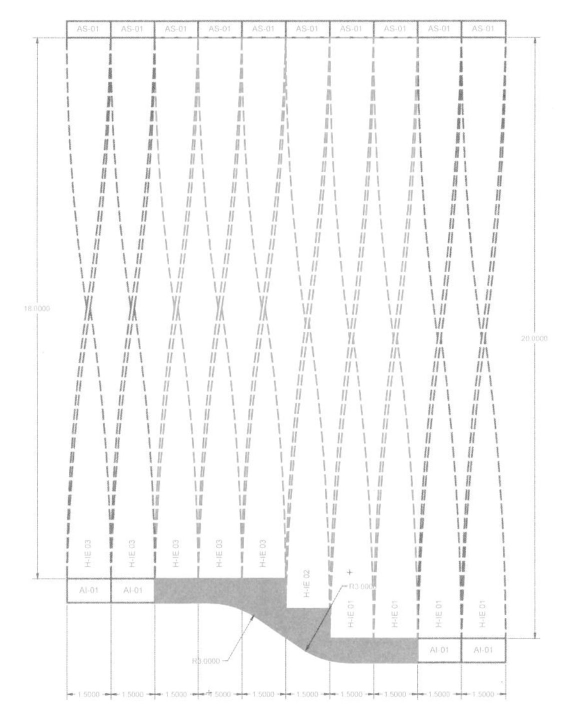

立面详图 FAÇADE DETAIL

Given Tabasco's tropical climate and its severe solar incidence and humidity levels, concrete was selected as the project's design material, a material both resistant and with extraordinary aging qualities. With the development of innovative construction technologies, the project would seek a new image for Liverpool. The result was a façade that's built by combining 5 different types of precast pieces shaped like a propeller. Each propeller rotates 180° on its axis, heights vary between 16 to 20 meters, depending on their position.

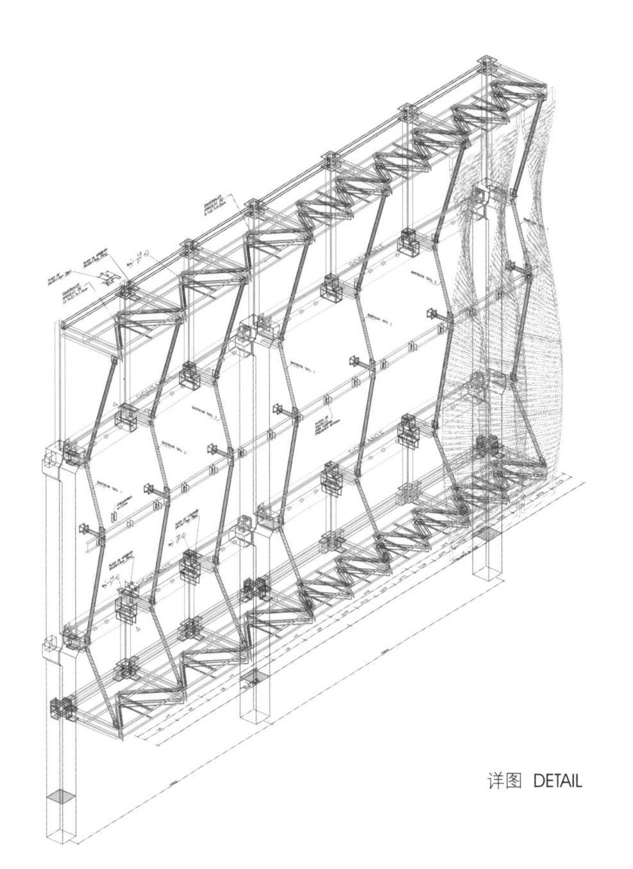

详图 DETAIL

这些简单可控的变化形式为每一个部件带来了无尽的视觉效果，整体来看又给人一种动态感。当人们从远处欣赏或驾车高速经过时观望，这种效果更为明显。近距离俯瞰，混凝土浆片就像是精致的木材，酸性面漆涂层不仅显现出了混凝土的纹理，而且为混凝土带来了奇异的外观。

These simple and controlled variations create numerous results for each piece, which as a whole, give a sense of movement, this effect is better appreciated at a distance and when passing trough by car at high speed. From up close, the concrete looks like a fine wood, the acid layer applied as a final coating, brings out the concrete's grain, which in return, gives the material this odd appearance.

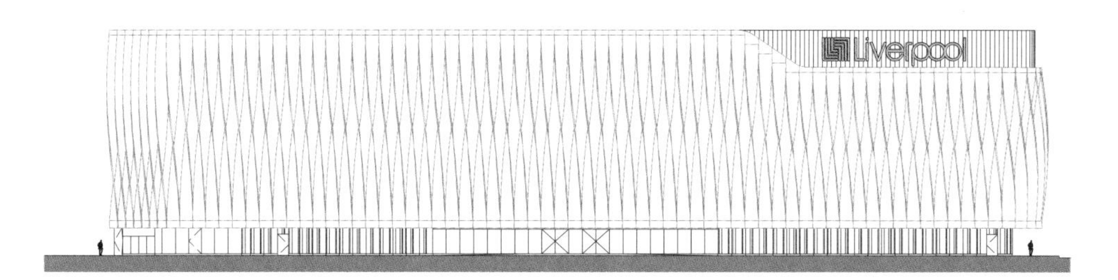

西北立面图 NORTHWEST ELEVATION

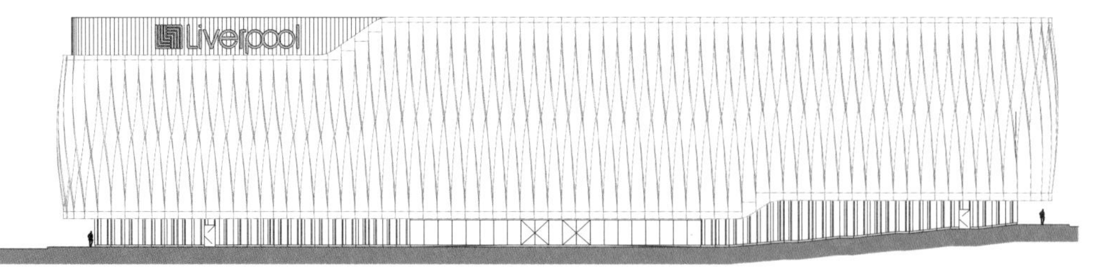

东北立面图 NORTHEAST ELEVATION

“剪裁制版”
"construction"

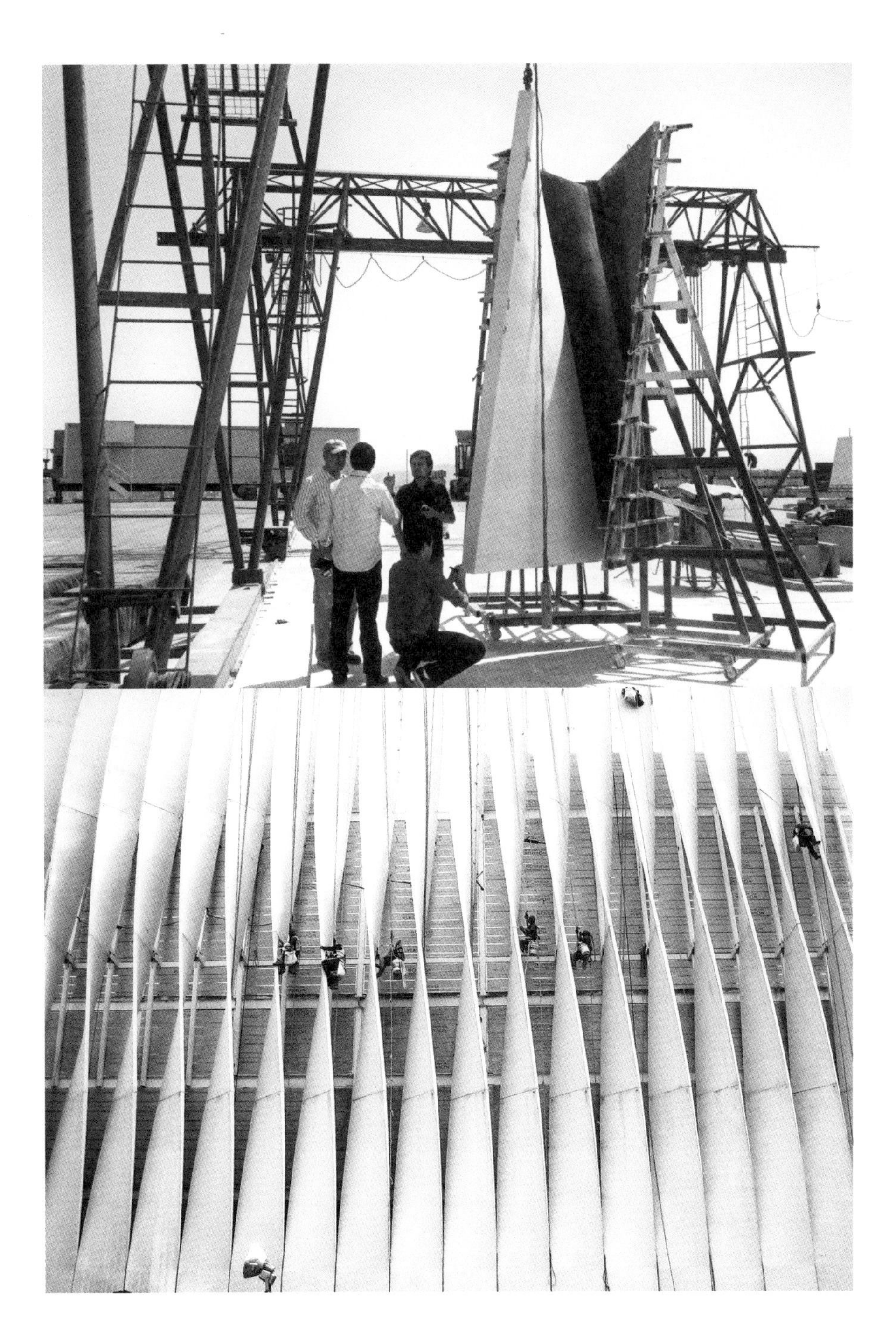

Additionally, the light changes that occur during the day, and the artificial lighting at night, provide an interesting mixture of colors, reflections and shadows, achieving an always changing and never static image for the façade.

另外，白天不断变化的日光及夜晚的人工照明为观者带来了色彩、反射和光影交叠的奇趣异景，形成了千变万化、永不静止的立面效果。

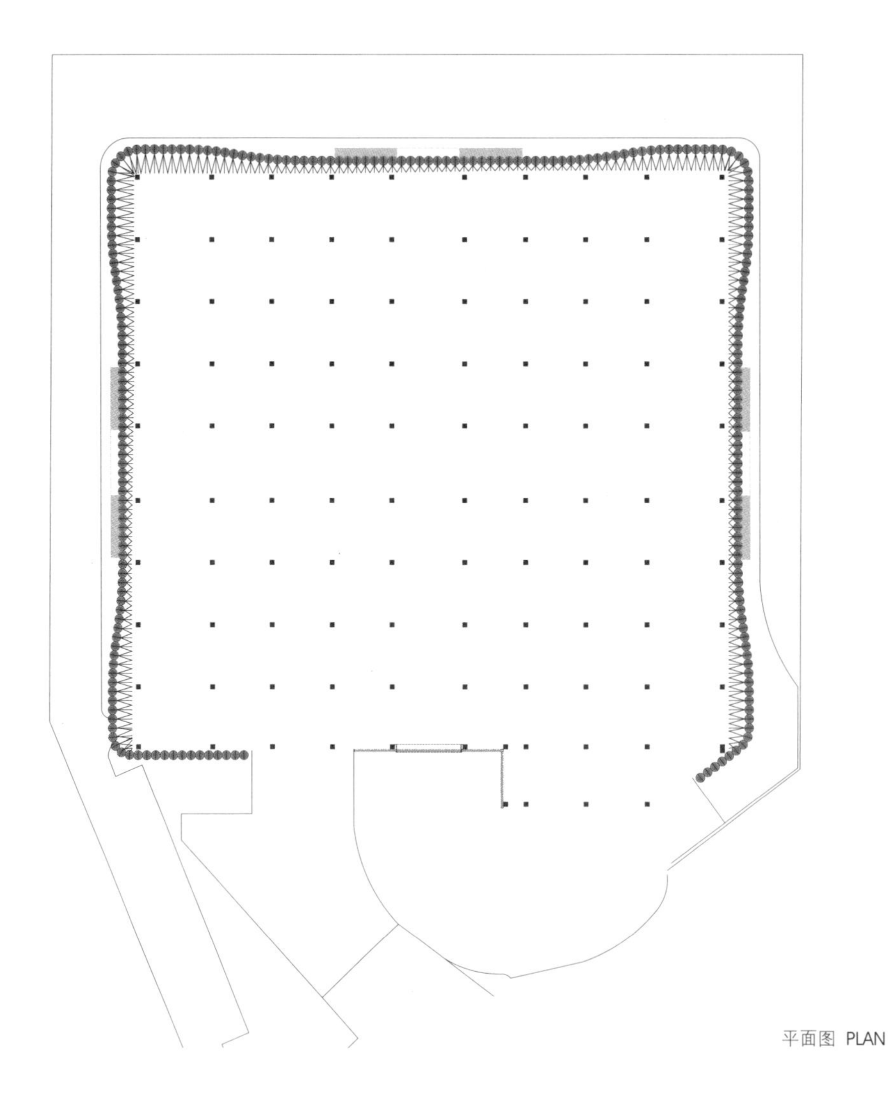

平面图 PLAN

Liverpool
20

建筑学院新门厅

NEW FOYER OF THE BUILDING ACADEMY

项目名称：

建筑学院新门厅

Project Name:

NEW FOYER OF THE BUILDING ACADEMY

项目设计： soma architecture
项目地点： 奥地利，萨尔茨堡
立面材料： 清水混凝土
竣工时间： 2012 年
摄影： F. Hafele

Architects: soma architecture
Location: Salzburg, Austria
Façade Material: exposed concrete
Completion: 2012
Photographs: F. Hafele

皮草一直是时尚界中高贵、典雅的象征。那柔软又奢华的感觉总令人爱不释手，蓬松而柔软的质地特别能给人以温暖的感觉。然后再用棕色皮质条带加以装饰，使服装的颜色与线条更加富有变化。新门厅的白色混凝土立面那连绵不断的折痕，仿佛风吹过后波浪起伏的毛皮表面一般，给人以变幻莫测之感。混凝土结构的中心用钢架支撑，锈蚀的表面不正是那皮质的装饰条带吗？这充满现代感与前卫感的建筑立面，给人们带来独特的视觉享受。

Fur and leather has always been a noble and elegant symbol in the fashion world. It fascinates people with the sense of luxury, and its fluffy and soft texture brings people a warm feeling. Decorated with brown leather belt, the colors and lines of the dress become more varied. The continuous creases on the concrete façade of the new foyer look like the undulating furs blown by the wind, giving people an unpredictable feeling. The center of the concrete structure is supported by a steel frame, whose rusty surface is just like the leather belt. This modern, avant-garde façade brings people a unique visual treat.

© 宋宁 (Bella Song)

"面料定制"
"Material"

soma 建筑事务所为建筑学院设计了新门厅和入口，建筑师将原工厂厂房改造为一处文化中心。经过改建，这个门厅将拥有多种功能——它既是学校的大厅，又可为相邻大厅举办文化活动时提供公共空间。

门厅的三维空间模式模拟了粒子流的形态。液体有三个基本参量：黏度、密度和表面张力。设计者用电脑对这三个物理参量进行了多种交互测试，设计出了带有多个孔洞、高度连贯的图案式样。

soma's design creates a new foyer and entrance topography to the Building Academy, and adapts an existing factory hall into a cultural venue. After the conversion, the foyer will serve multiple purposes - as a lobby for the school and as a public space for various cultural events in the adjacent halls.

The three-dimensional pattern was generated in a simulation of fluids based on particle flows. Liquids have three essential parameters: viscosity, density and surface tension. The interactions between these three physical properties have been tested on the computer in a series of variations to generate a pattern with a big amount of holes and a high level of coherence.

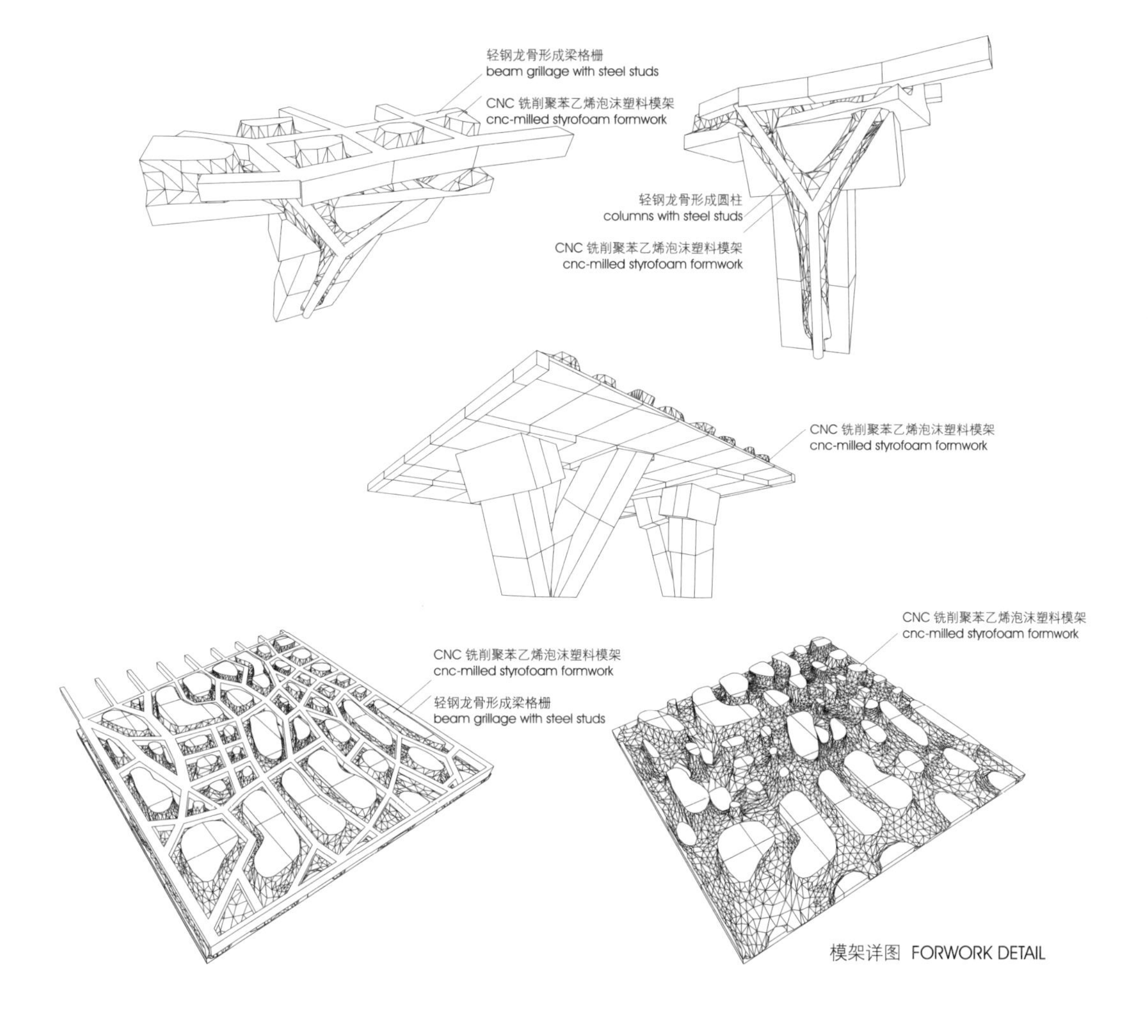

模架详图 FORWORK DETAIL

新的屋顶结构使建筑内外实现了连贯而流畅的过渡。它将各个不同的功能区连接成一个流畅的空间，营造出开放而怀旧的氛围。结构性能、照明、功能分区以及直觉引导系统相互关联，共同组成一个连贯结构。屋顶从一个框定的架构扩展成了一个自由伸展的几何体，这是对原空间特征的呼应。

The new roof structure creates a consistent and smooth transition between the exterior and the interior of the building. It connects the different functional areas in one fluid space and creates an open and evocative atmosphere. Structural performance, lighting and functional zoning, as well as the creation of an intuitively understandable guidance system correlate in one continuous structure. The widening of the roof geometry from a framed into a freely spreading geometry can be read as a contingent affiliation of the existing spatial characteristics.

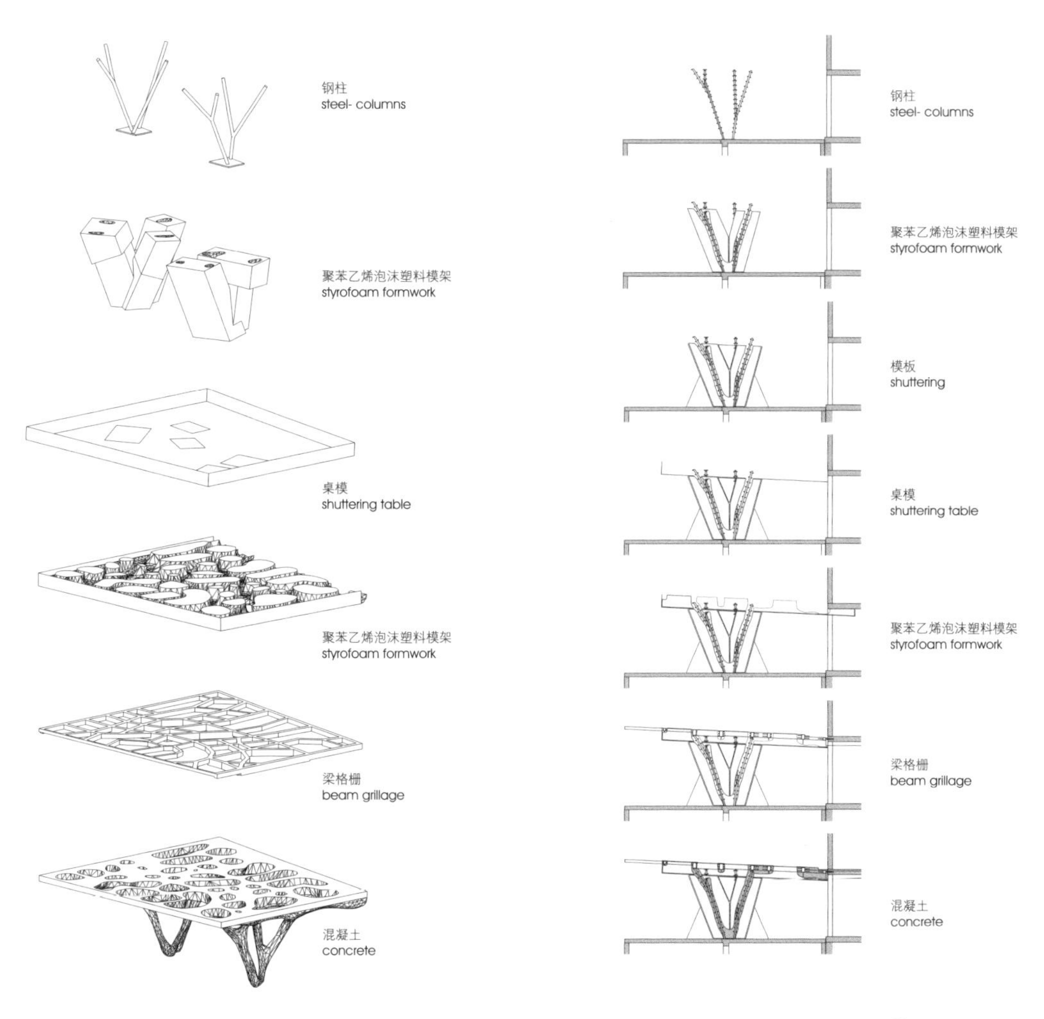

组装图 ASSEMBLY

这座露石混凝土复合结构体在框架构筑方面向设计者提出了严峻挑战。它的建成历经多次 1:1 建模的测试和优化。为了达到奇妙的光影效果，这座混凝土结构的三角形表面经过多次铣削工艺实验。屋顶精密的几何造型和创新性的建筑过程，反映出建筑学院在混凝土应用领域中的实力和教学特色。

Realizing this complex structure in exposed concrete posed major challenges in the area of formwork construction, which was tested and optimized in 1:1 mock ups. The finely triangulated surface of the concrete structure allows a subtle play of light and shadow and was achieved by testing different milling techniques. The advanced geometry of the roof and its innovative construction process reflect the competencies and teaching focus of the Building Academy in the field of concrete application.

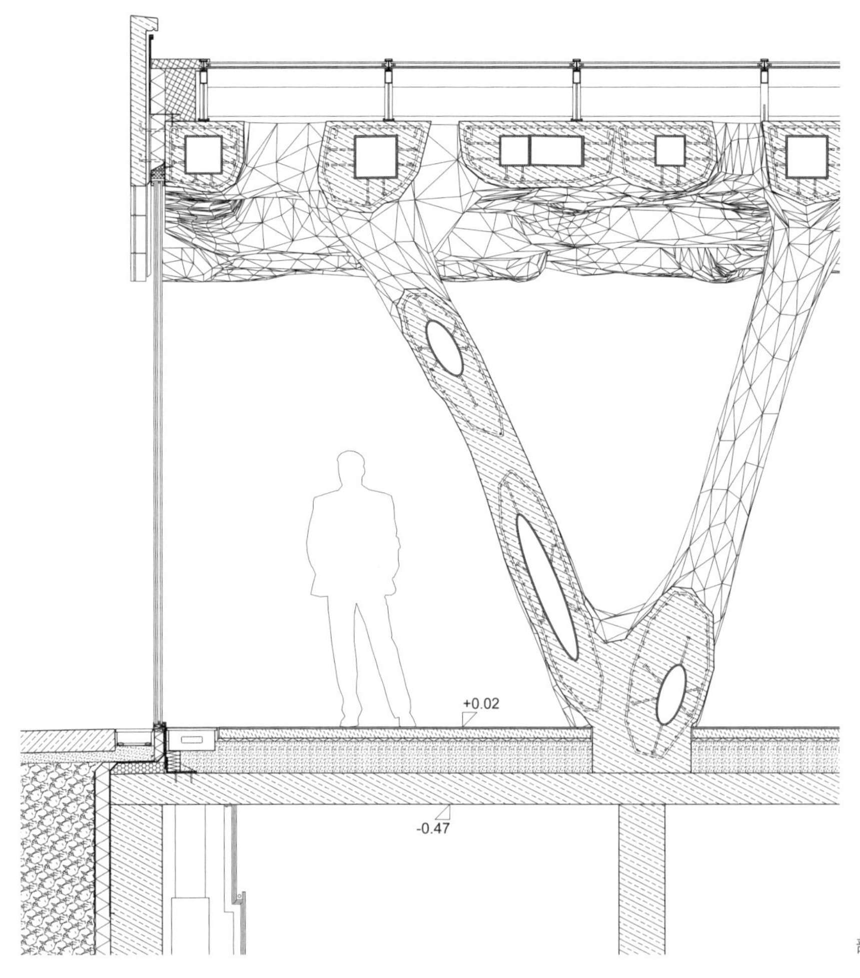

剖面图 SECTION

海门林纳档案馆

HÄMEENLINNA PROVINCIAL ARCHIVE

项目名称：
海门林纳档案馆
Project Name:
HÄMEENLINNA PROVINCIAL ARCHIVE

项目设计： Heikkinen-Komonen Architects
项目地点： 芬兰，海门林纳
立面材料： 混凝土
摄影： Jussi Tiainen

Architects: Heikkinen-Komonen Architects
Location: Hämeenlinna, Finland
Façade Material: concrete
Photographs: Jussi Tiainen

把文字印到衣服上的做法早已有之。基本上疯人印"疯言"，狂人写"狂语"，恋人秀"甜蜜"，文青有点"小文艺"，连不识字的大妈也穿着缀满外文的T恤。档案馆该"穿"什么"衣服"？当然是缀满神秘密码般的"天书"装最适合了。海门林纳档案馆是芬兰历史的重要宝库，其收藏的文件可追溯到16世纪。该档案馆的外墙由大型蚀刻混凝土面板构成，刻有从历史文献中摘录的图案、文字和文本片段，与建筑功能相呼应。

It has been long to use letters on clothing as decoration. People can choose different words on the clothes to show their personality, and some people wear T-shirt with foreign words that they couldn't read. What kind of clothes is suitable for an archive building? A garment with mysterious hieroglyphics all over is the best choice. Housing documents and texts dating back to the 16th century, the Hämeenlinna Provincial Archive is an important repository of Finnish historical information. Its exterior walls are constructed of large etched concrete panels displaying images, illuminated letters and fragments of text extracted from the historical documents, producing a façade that prominently heralds the building's purpose.

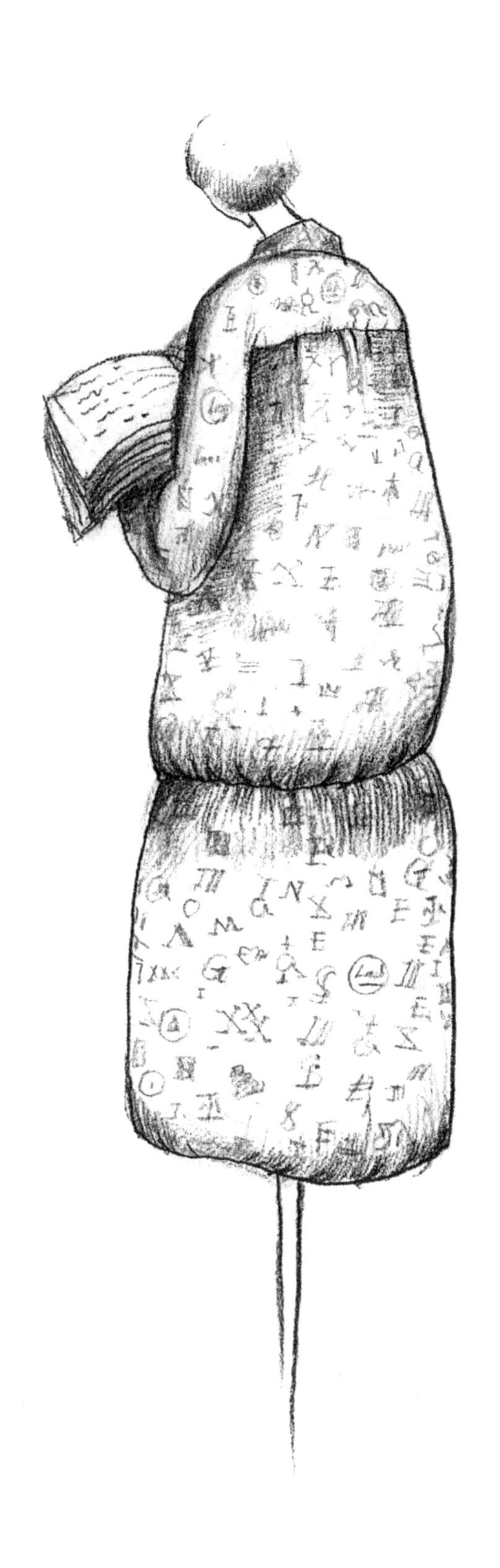

© 郑亚男（Nancy Zheng）

图书馆立面采用绘图混凝土构筑手法，在芬兰尚属首创。这一技术基于这样的原理：将一种特制的薄膜铺展在印有图案的模具桌上，上面再敷上一层表层混凝剂；经过腐蚀，浅色的图案就在黑色的混凝土表面显现了出来。

The graphic concrete method employed has been patented in Finland. The technology is based on applying a surface retarder to a special membrane that is spread over the mould table, on which the desired patterns are printed. After corrosion, these appear light against the dark surface concrete.

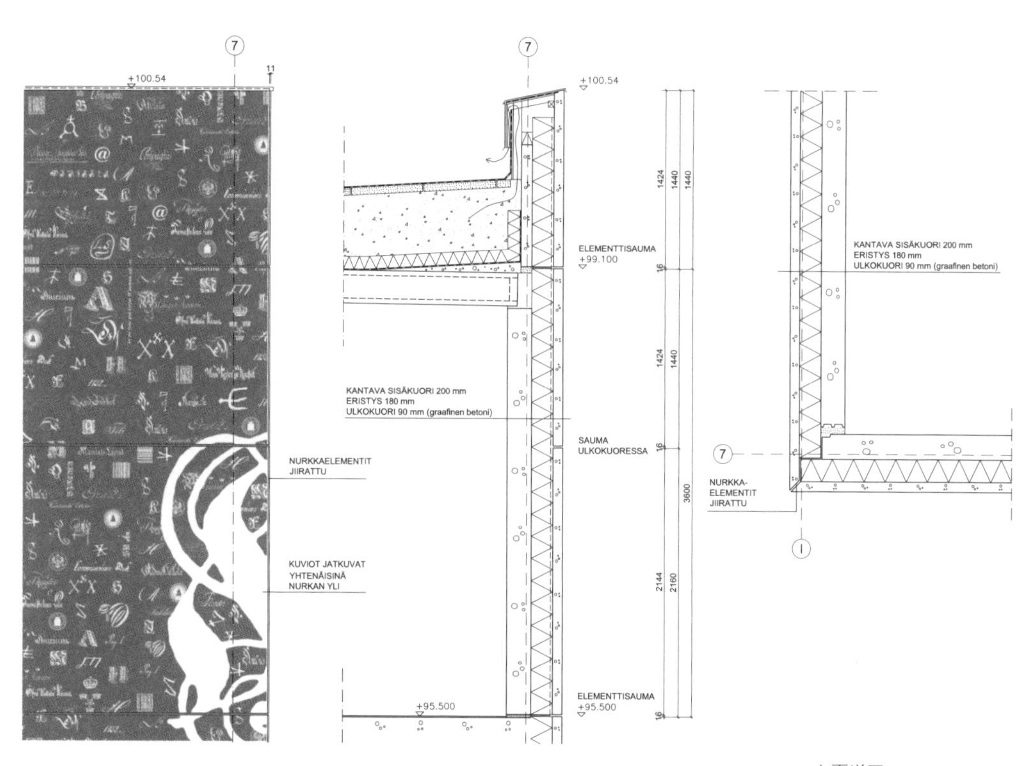

立面详图 FAÇADE DETAIL

底层开阔的场地内，各个不同功能空间布局如群岛一般，自习室、礼堂、图书馆、自助餐厅和展览空间等分布其中。这个透明而缤纷的楼层上面是档案馆三层高的盒状立方体体量，刻有图案的混凝土覆盖着其表皮和内墙。在庭院一侧，办公室和工作室覆盖着棕色的铝板，一条带有天窗的走廊将其与建筑的其他部分分隔开来。

The open area on the ground floor is organized as an archipelago of different spaces like study rooms, auditorium, library, cafeteria and exhibition space. Above this transparent and colorful ground floor is three-floor high solid archive box, which is covered with graphic concrete elements both on the exterior and interior side. Offices and workshops in the yard side are covered with brown aluminum plates and they are separated from other parts of the building by a skylight canyon.

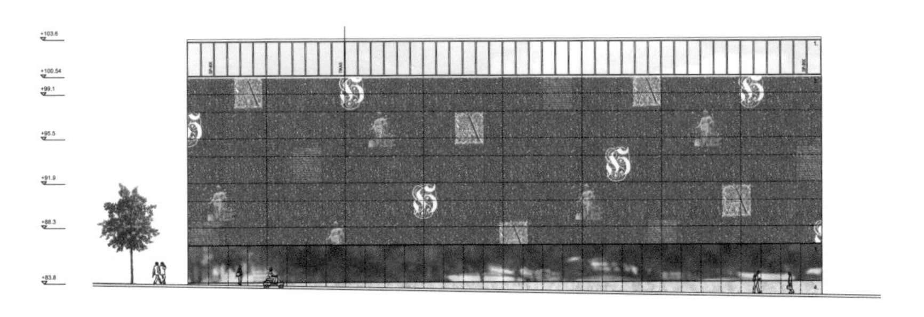

西立面图 WEST FAÇADE

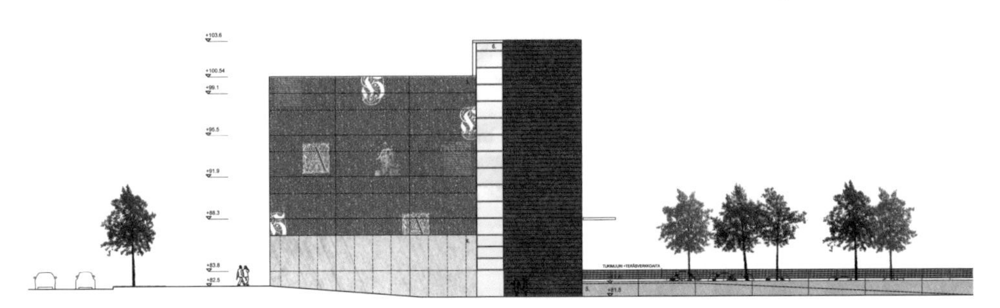

西南立面图 SOUTHWEST FAÇADE

东北立面图 NORTHEAST FAÇADE

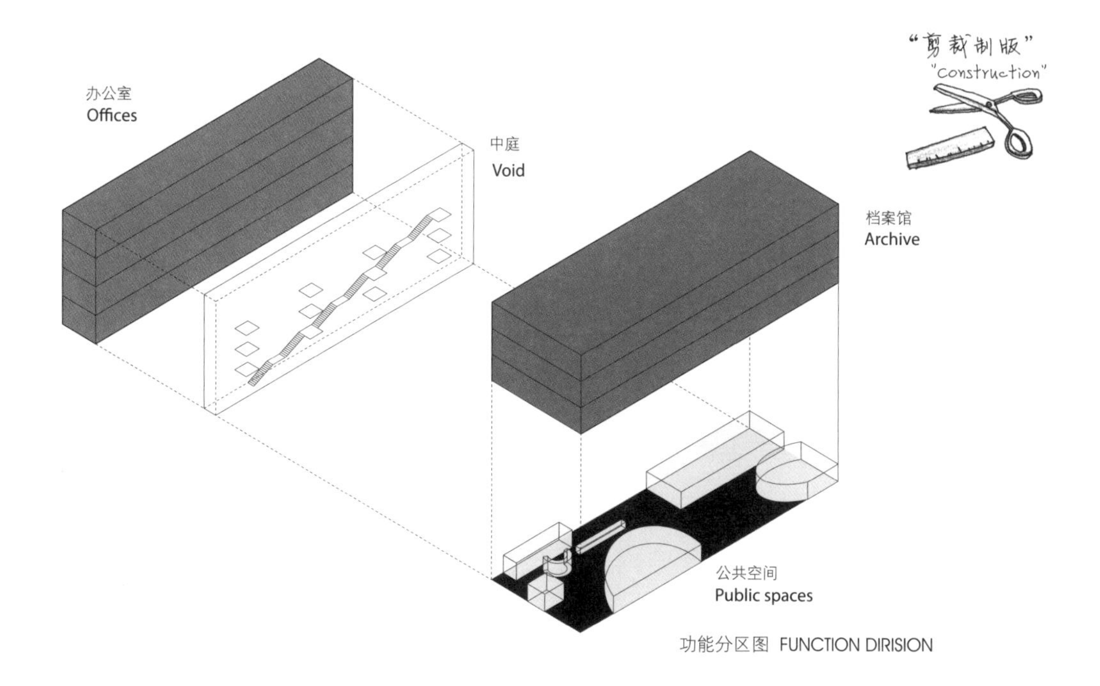

功能分区图 FUNCTION DIRISION

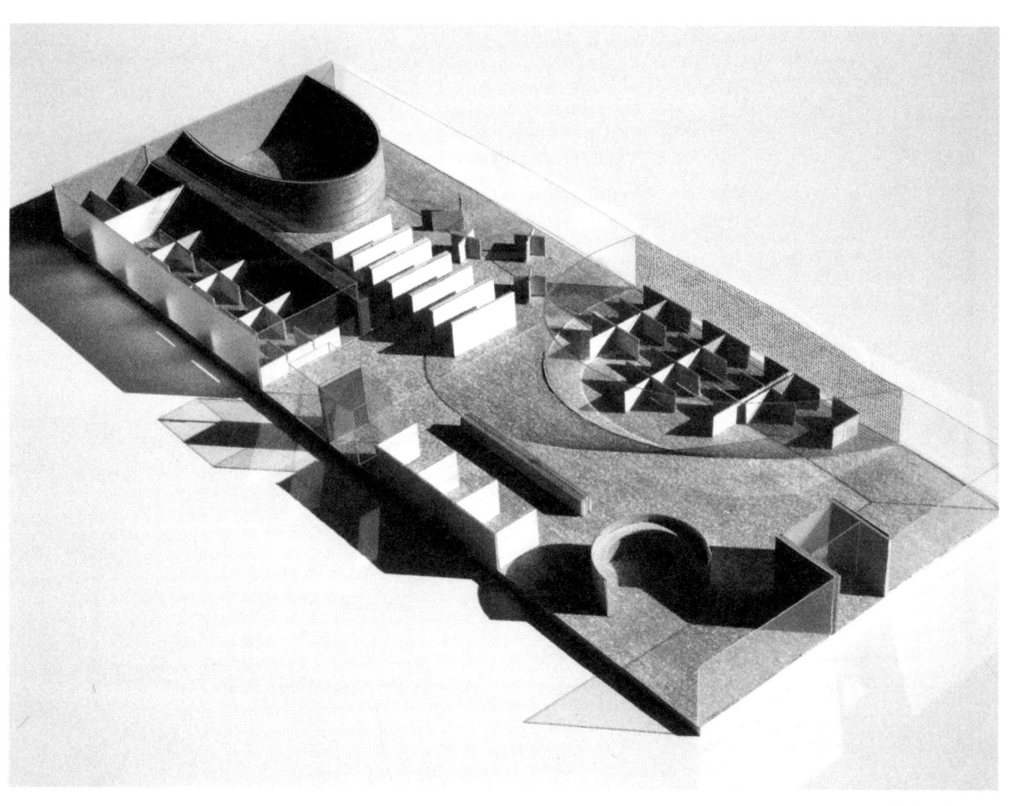

模型 MODELS

这座省档案馆不仅是一个历史文献的仓库，还是一个重要的公共机构。它保存和陈列着集体和个人的往昔，并供研究人员查询使用。一座重要的公共建筑应当在城市景观中醒目且易于辨识。作为我们共同记忆的储藏所，它理应在外观上与平淡无奇的仓库或办公楼区别开来。

The Provincial Archive is not only a store for old documents, but a significant public institution that preserves, displays and makes accessible to researchers the records of the collective and private past. It is an important public building that should play a visible and recognisable role in the townscape. As the depository of our collective memory it should not look like an anonymous warehouse or office block.

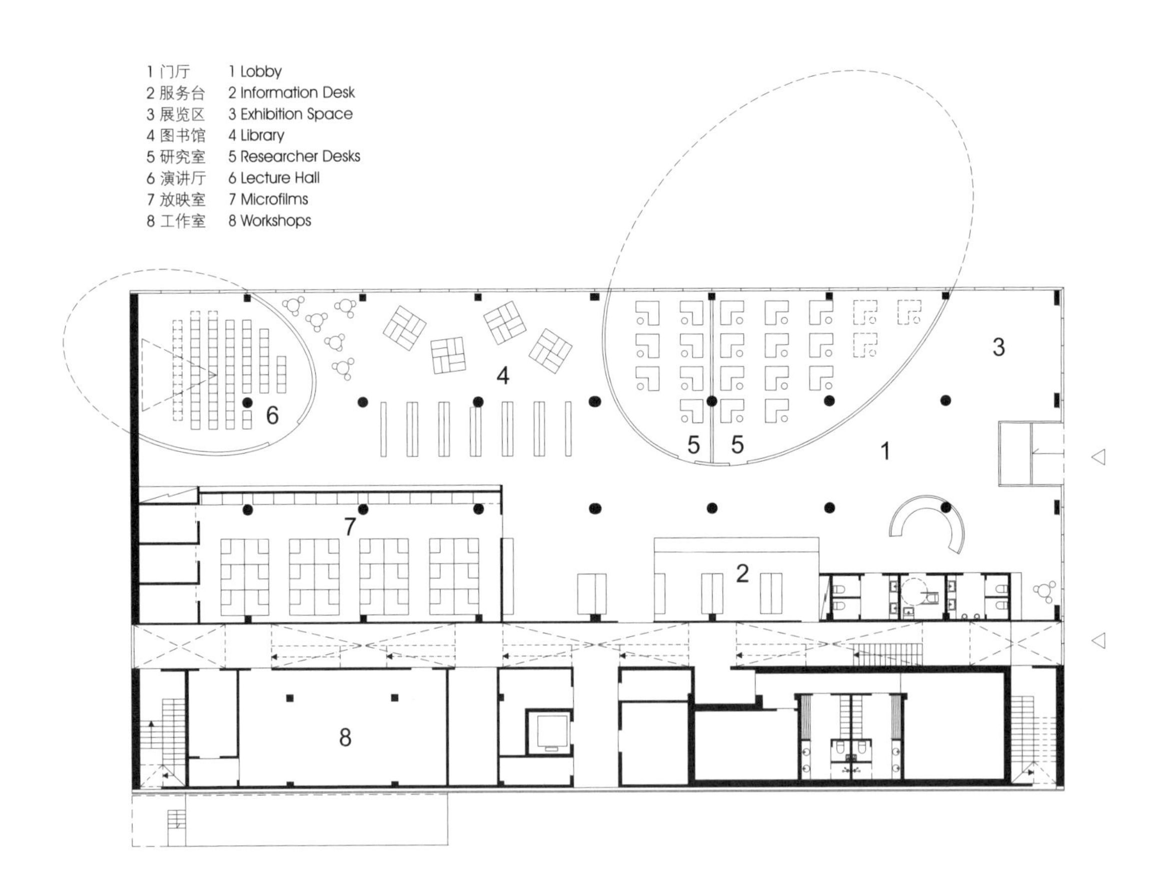

平面图 PLAN

两百年纪念市政中心

BICENTENNIAL CIVIC CENTER

项目名称：

两百年纪念市政中心

Project Name:

BICENTENNIAL CIVIC CENTER

项目设计： GGMPU Arquitectos + Lucio Morini
项目地点： 阿根廷，科尔多瓦
立面材料： 混凝土
竣工时间： 2012 年
摄影： Claudio Manzoni, GGMPU + Lucio Morini

Architects: GGMPU Arquitectos + Lucio Morini
Location: Córdoba, Argentina
Façade Material: concrete
Completion: 2012
Photographs: Claudio Manzoni, GGMPU + Lucio Morini

镂空的服装总是带有一点妩媚与性感，但却不张扬；能带给人一丝遐想，却不是浮想联翩。镂空的外衣配上深色吊带裙是绝妙的服装组合，菱形镂空图案中肌肤若隐若现，透露出优雅的性感。穿上这样一件衣服，平添不少女人的柔美与婉约。正如这座市政中心由混凝土构筑成的外壳，菱形的镂空图案交织排列开来，仿佛给建筑穿了一件菱形镂空的“外套”一样。在遮挡强烈日晒的同时，又起到了美化建筑的作用。

The clothes with hollow-out elements always give a feeling of charming and sexy, but not too showy, and bring people beautiful reveries. The hollow-out smock and the dark suspender skirt make a perfect match. The hollow-out in rhombus patterns exposes a little bit skin indistinctly, showing a feel of elegantly sexy. Dressed in it, the softness and grace of women are fully presented. Bicentennial Civic Center is covered with a concrete shell with rhombus hollow-out pattern. The shell acts as a smock for the building, which not only can block strong sunlight but also embellish the building.

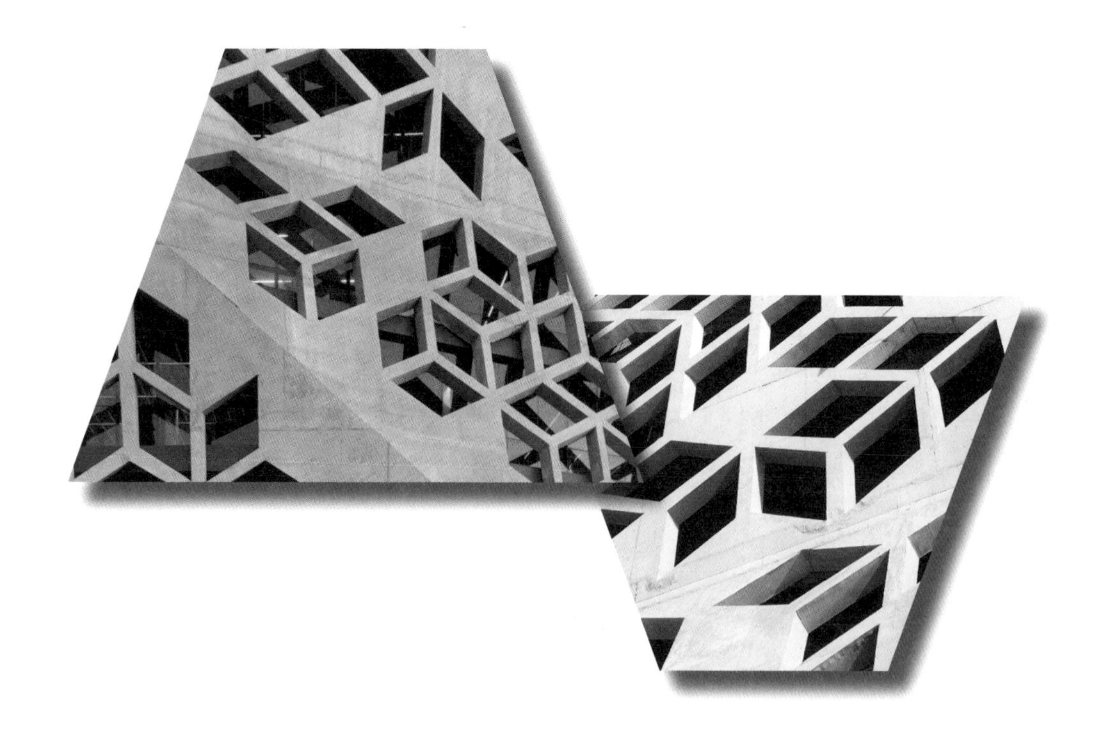

© 宋宁（Bella Song）

市政中心由 A、B 两座建筑构成。A 楼是一座高度中等、由混凝土构筑的多面棱柱形建筑，内部容纳了政府主要部门。建筑是一个 45 m 高的长方体，基底边棱长 26 m，并在 16 m 的高度进行了 20° 的折转。这种折转使楼体形成了以三角形为基面的更为复杂的体态，制造出了极为奇异的光影效果。为了避免折转产生的三角形立面沦为枯燥的平板，设计者在表皮做出了一系列镂空菱形组合图案，为各个立面增添了三维效果。

The Civic Center consists of two buildings, A and B. Building A is a medium-rise building housing the ministries dominates the complex: a faceted prism made out of concrete. It is based on a square cuboid 45 meters high and with a side measuring 26 meters, which at a height of 16 meters suffers a 20-degree rotation. This movement generates a more complex morphology based on triangles, which in turn produces a very particular play of shadows and light. To avoid the flat character present in the early studies for the triangular façades produced by the rotation, a series of geometric rhomboid-based combinations were explored in order to instill tri-dimensionality to the very planes of the façades.

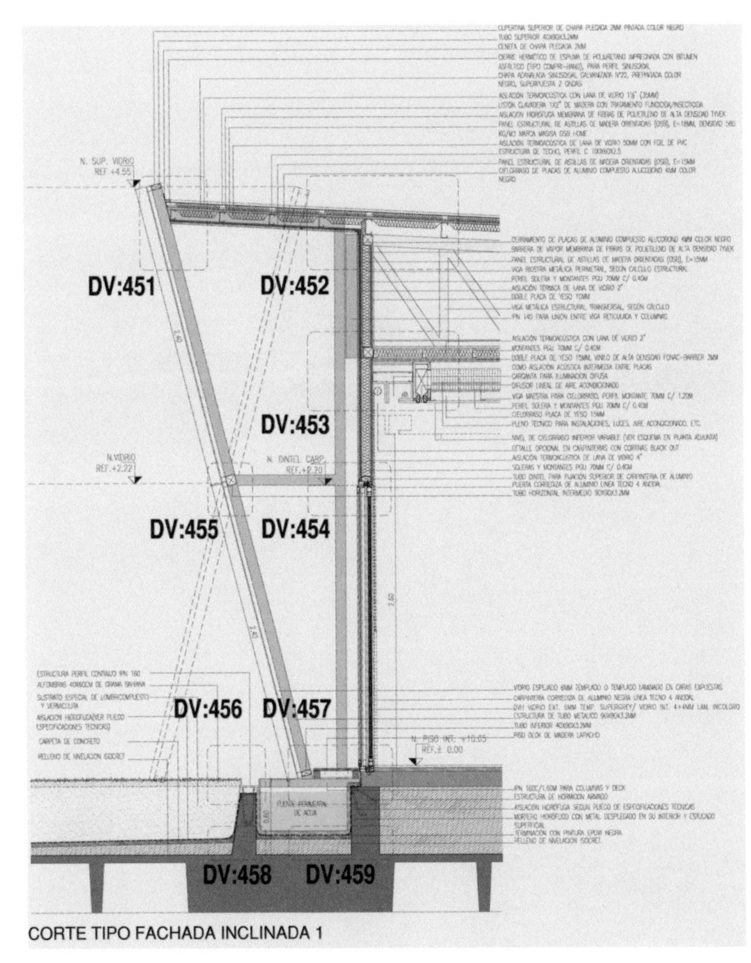

CORTE TIPO FACHADA INCLINADA 1

详图 DETAIL

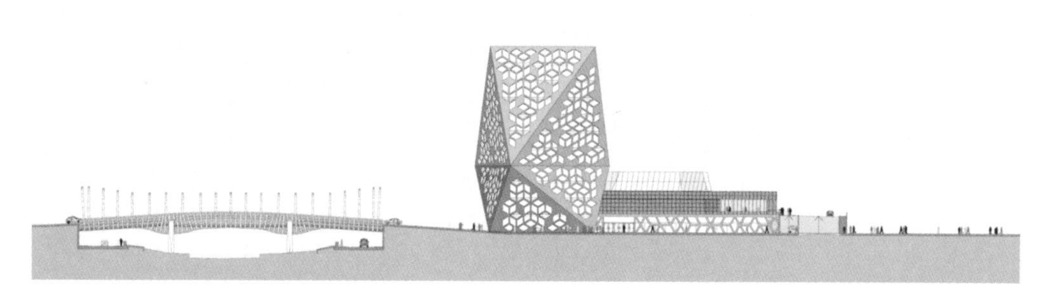

北立面图 NORTH FAÇADE

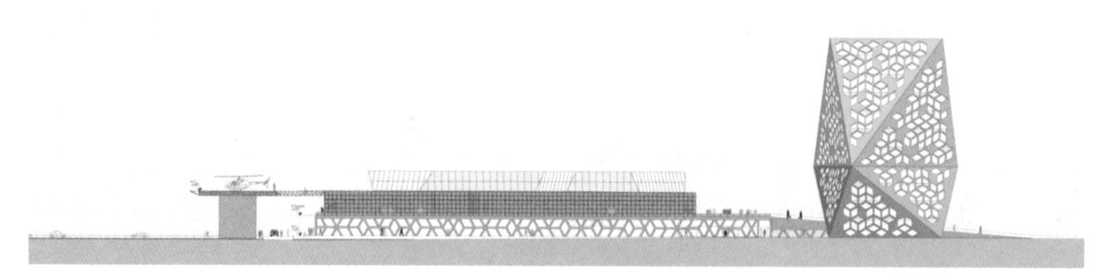

东立面图 EAST FAÇADE

与 A 楼形成对比，B 楼是一座水平伸展的建筑，容纳了各种行政功能。其下方是一个混凝土基底，包含了停车场和用于举办交际、社会活动的区域。其表皮与 A 楼相呼应，也采用了菱形元素，但尺寸有所变化。大楼上方是一个被金属网所包绕的绿色空间，可供落叶蔓藤植物自由攀爬。植物有助于调节温度，冬日让阳光投进室内，夏天又能够遮挡强光。同时建筑还拥有了一个种满了花草的绿色屋顶。

In contrast with Building A, Building B is a horizontal slab and houses administrative functions. It is placed over a concrete base incorporating parking and areas that foster interrelationships and social events. Its language is related to that of Building A, including the rhomboids but changing their scale. On top of this rests a green volume, which is enclosed by a metal mesh enabling deciduous vines to climb freely. The vines help regulate temperature, allowing sunshine to reach the building in winter while sheltering it in the summer. The roof is also green and is covered by grass and flowers.

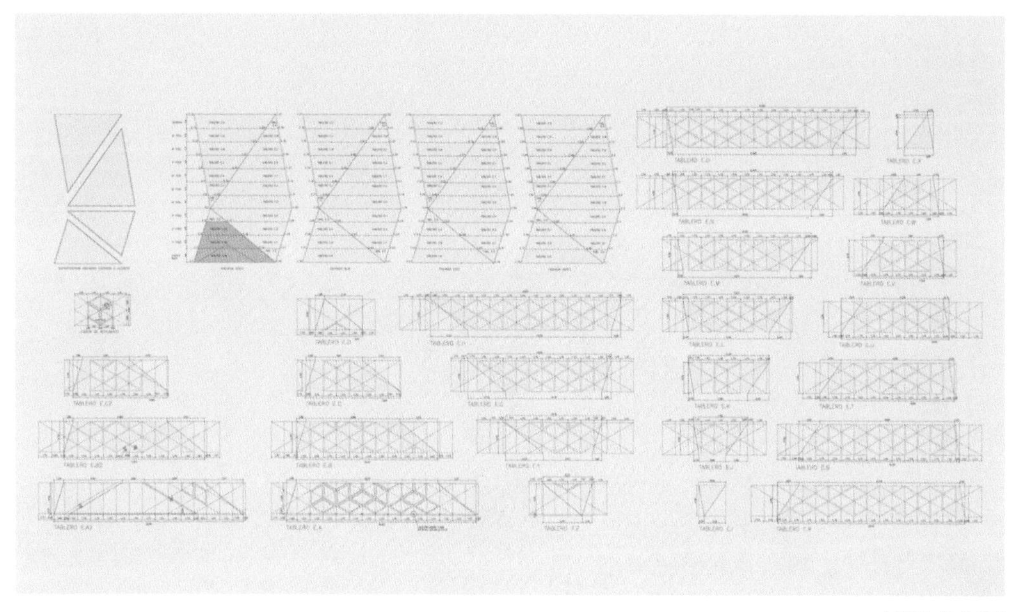

立面图 FAÇADE

该项目是阿根廷科尔多瓦省第一座专为两百年纪念而设计的行政综合体。它位于城市历史中心的边缘，那里曾经铺设着铁路轨道。这座城市新建筑有利于改变这片河流区域落后的面貌，将整个地块转变为一个新的城市中心。

The Bicentennial Civic Center is the first administrative complex to have been designed specifically for that end by the Province of Córdoba, Argentina. It is located on the edge of the historic center of the city, on a lot that used to belong to the railway tracks. This new urban operation will help diminish the image of the river as backyard of the city, thus transforming the whole sector into a new downtown front.

1 大厅
2 接待处
3 门厅
4 视听室
5 预备室
6 保健室
7 残障人士保健室
8 衣帽间
9 前室
10EPEC 控制室
11 单人间及变电室
12 发电机组
13 操作室
14 空调设备
15 垃圾存放间
16 管道间
17 储藏室
18 设备间
19 技术间
20 监控室
21 库房
22 政府代表公寓入口
23 办公室
24 开放办公室
25 酒吧
26 厨房
27 休息区
28 后勤部
29 演奏厅另一入口
30 集市

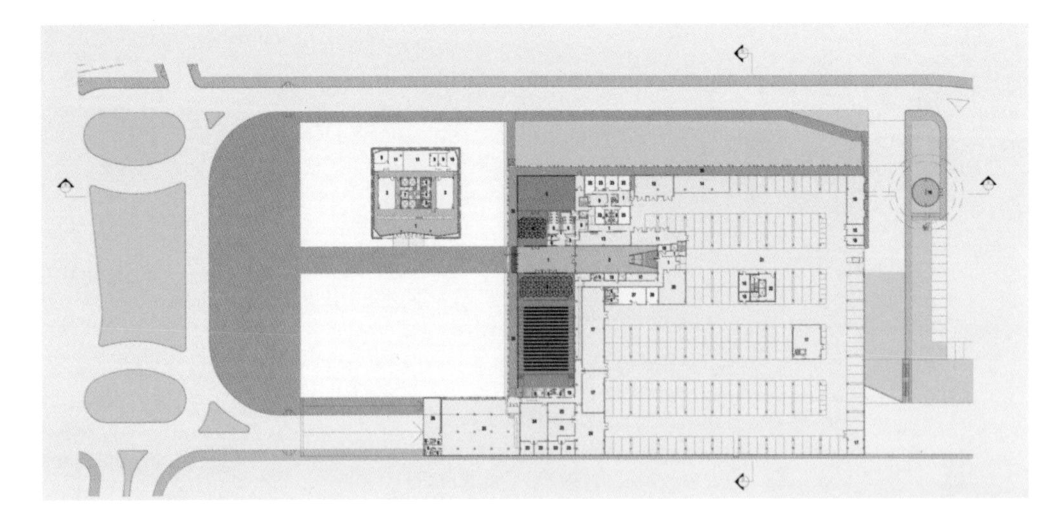

一层平面图 GROUND FLOOR PLAN

蒂艾 RATP 公共汽车站
RATP BUS CENTER IN THIAIS

项目名称：

蒂艾 RATP 公共汽车站

Project Name:

RATP BUS CENTER IN THIAIS

项目设计： ECDM architectes
项目地点： 法国，蒂艾
立面材料： 混凝土
摄影： Benoit Fougeirol, Philippe Ruault

Architects: ECDM architectes
Location: Thiais, France
Façade Material: concrete
Photographs: Benoit Fougeirol, Philippe Ruault

黑色短外套一直是时尚潮人的必备单品，既保暖又百搭。但若是纯黑色未免有些单调，以深色网纹装饰则更显优雅大方，再于腰间系一条彩色腰带，勾勒出女人优美的曲线。不但打破了冬季的沉闷，又平添了几分春意盎然。公共汽车站那沉稳的黑色混凝土外立面，用圆形凸起做装饰，让深暗的立面在阳光的照射下，呈现出黑、灰交织的效果，深浅不一、凹凸有致。搭配色彩斑斓的大块玻璃窗，使建筑立面更加绚烂，展现出勃勃生机。

A small Black coat is a necessity for fashion girls. It is not only warm but can go with any clothes. However, it would be monotonous if it is all black, and the dark mesh patterns make it more elegant. Moreover, a colorful bell at the waist is better to articulate the beautiful curves of women. It not only breaks up the winter dullness, but also brings out some spring warm and vigor. The black concrete façade of RATP Bus Center is comprised of numerous round embossments. In the sunlight, black and gray are interwoven on the dark façade, presenting an image of different brightness and depth. Large colorful glass windows further glorify the façade and fill the building with vitality.

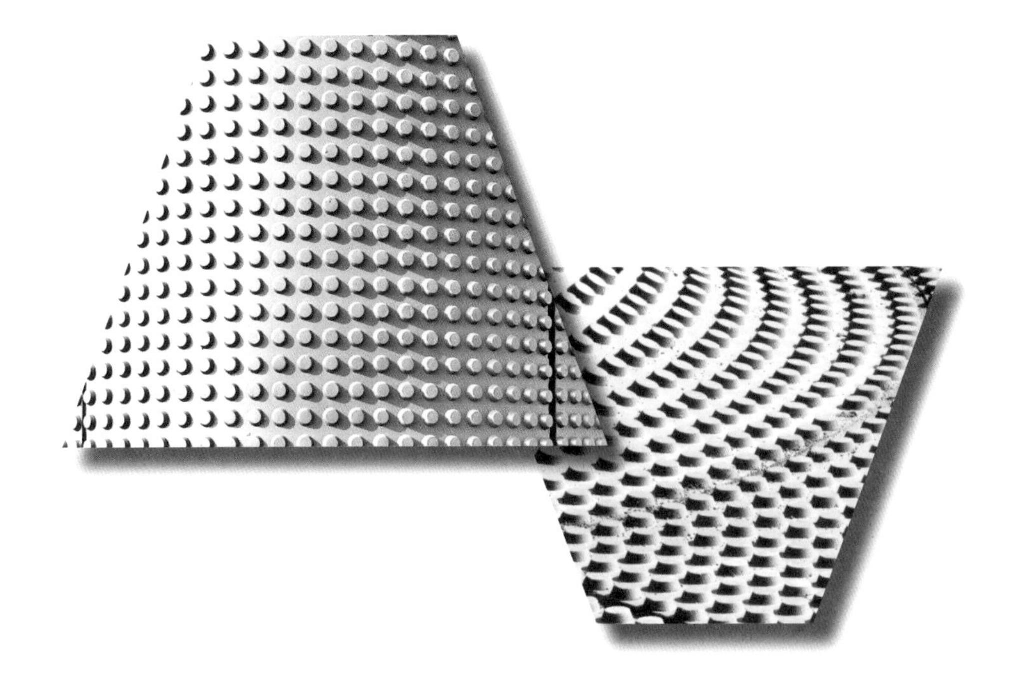

"概念创意"
"Concept"

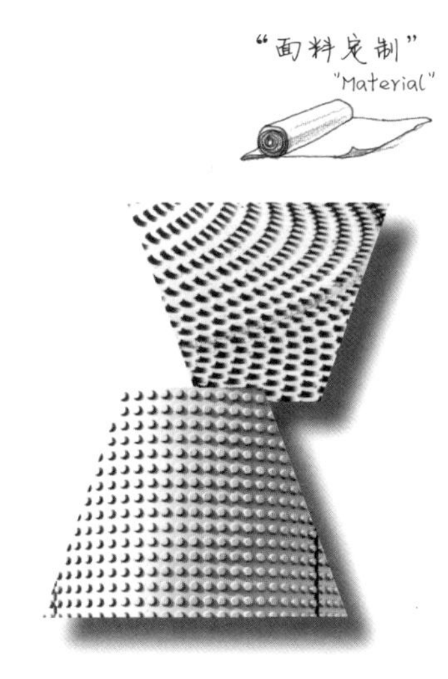

建筑师首先改造了地面使其变形，并将原混凝土板进行了扭曲处理，然后上方续接了一种相似的材料 Ductal®。这也是一种混凝土材料，但视觉效果更佳炫目。它满足了建筑各种复杂的需求，如结构的灵活性、规划的不断演化、并能通过模具的设计来设定其精度、密度和均匀性。它保证了从街道、地面到建筑立面、顶棚及屋顶平台的连贯性，使建筑整体没有任何缝隙。

The building starts with the deformation of the ground, the distortion of the existing concrete slab, and continues it with an apparently similar material Ductal®, still concrete but a dazzling sheet of concrete, which responds to very sophisticated demands: informality of the structure, constant evolution of the plans, dematerialization, precision, density, homogeneity of aspect according to the mould designed. It ensures a continuity of the ground from the road, to the skin of the façades, the suspended ceilings and the terrace rooftop without any rupture.

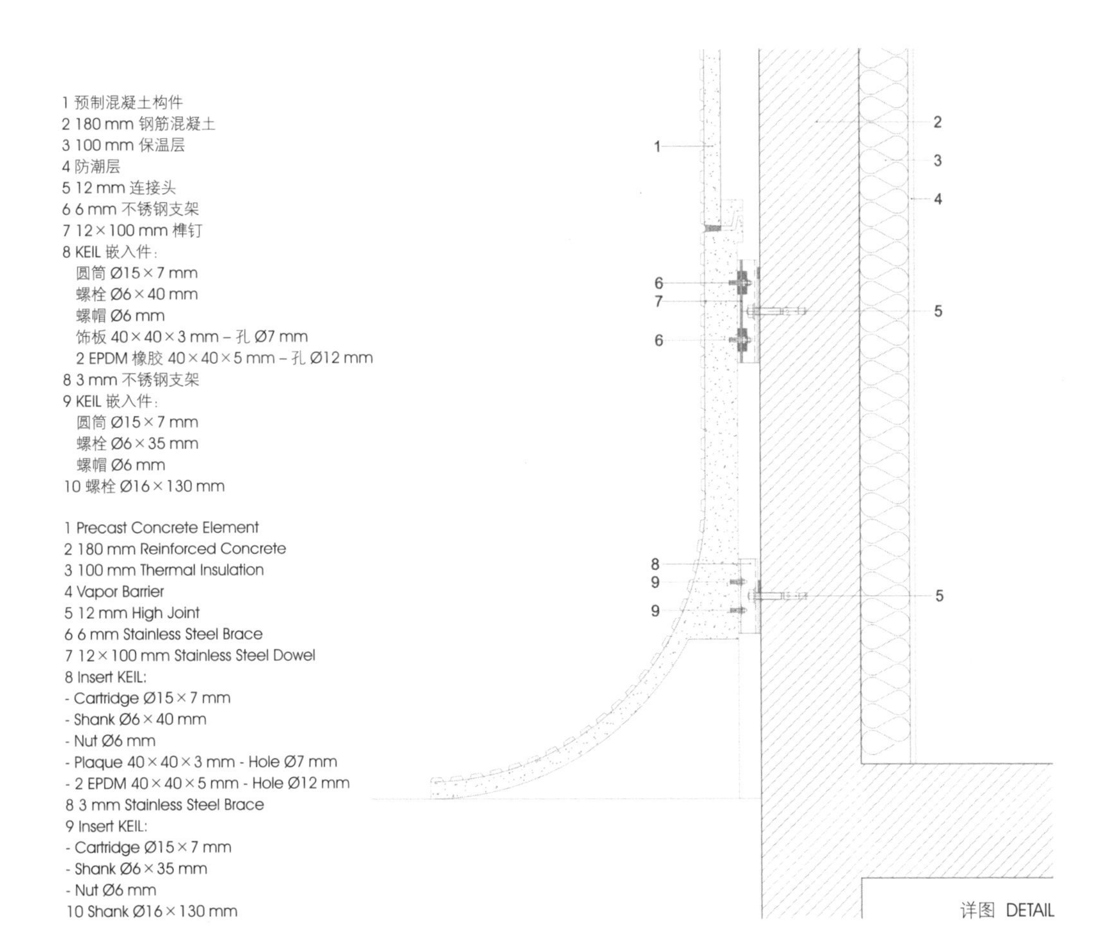

详图 DETAIL

建筑融入了这种材料的结构品质，因而看起来既没有起始也没有终端。建筑拥有一个连贯的表皮，这使人们难以界定其边缘所在，就像是一整块弯折的混凝土板，在整个建筑场地伸展延绵。整座建筑看起来好像是一个圆角的完整体块，光滑闪亮。建筑占地面积小，只有 35 m x 35 m，共两层，是一座高密度建筑。它缺少活力，但充满宁静与神秘，就像“摩尔曼斯克水面上的库尔斯克号核潜艇的外壳”。

Combined with the structural qualities of the material, the building has neither a beginning nor an end. It is a continuity of a surface of which we can, depending on what you aim at, not control the limits. It is an inflection of a slab which spreads on the whole site. The building appears like a monolith with rounded edges, polished somehow. Characterized by a dense square plan (35m x 35m) developed on 2 levels, it presents itself as a dense building, inert, deaf, and enigmatic as "the hull of a Russian submarine in the waters of Murmansk".

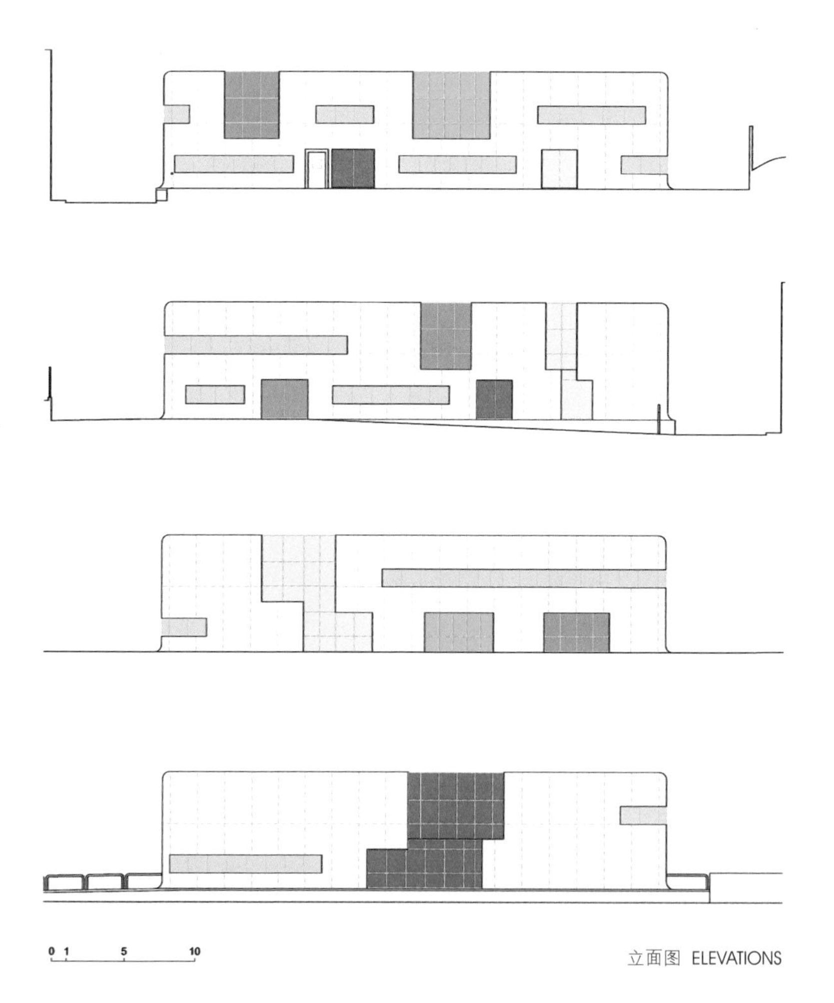

立面图 ELEVATIONS

"剪裁制版"
"Construction"

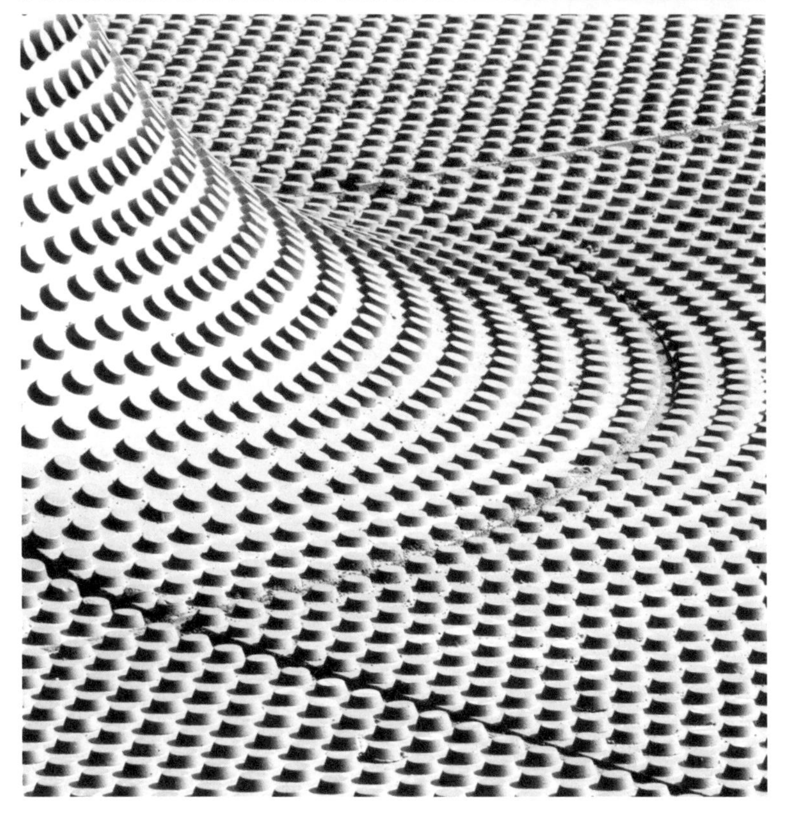

材料的灵活性和精确的塑形特性使建筑的纹理极为丰富。Ductal® 材料拥有良好的抛光性和精确性，赋予了混凝土表皮精密度和连贯性，实现了建筑外观的完美衔接。建筑表皮厚 3 cm，呈单一纹理，布满了浮雕般的圆点，就像 LEGO 积木一样。"窗子"是用切割刀片切割出来的，形成凹陷的体量，露出混凝土薄壳下方的彩色玻璃。窗子上布满银色的镜面圆点，并拥有缤纷的色彩，其灵感来源于主干道边缘办公楼的彩色幕墙。

The texture games are facilitated by the flexibility of the material and its ability to be molded with precision. The quality of finish and rigor of Ductal ® contribute to ensuring this concrete skin, accurate, continuous, perfect connections. The skin, 3 cm thick displays a single texture of dots in relief like a game of "LEGO". The "windows" are cut with a cutter blade, surgical incisions generating volumes in negative which reveal colored mirrors under the thin crust of concrete. Treated with silver mirror dots, chromatics of the glass products are inspired by the tinted curtain wall fronages of the office buildings which border the main road.

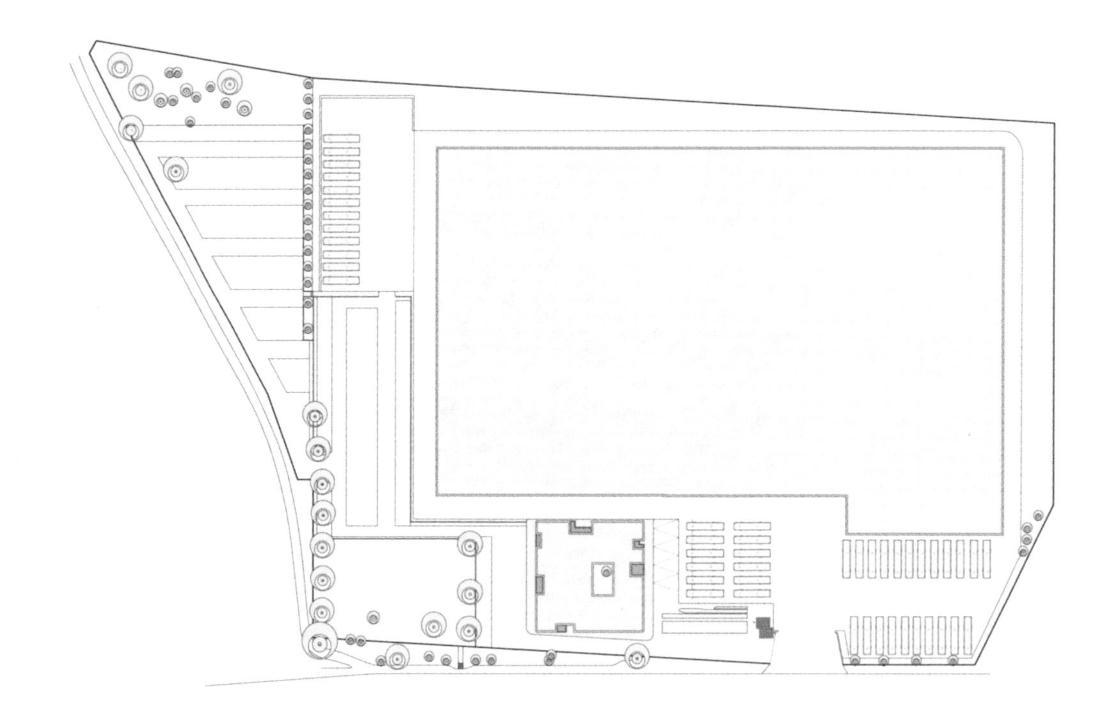

平面图 PLAN

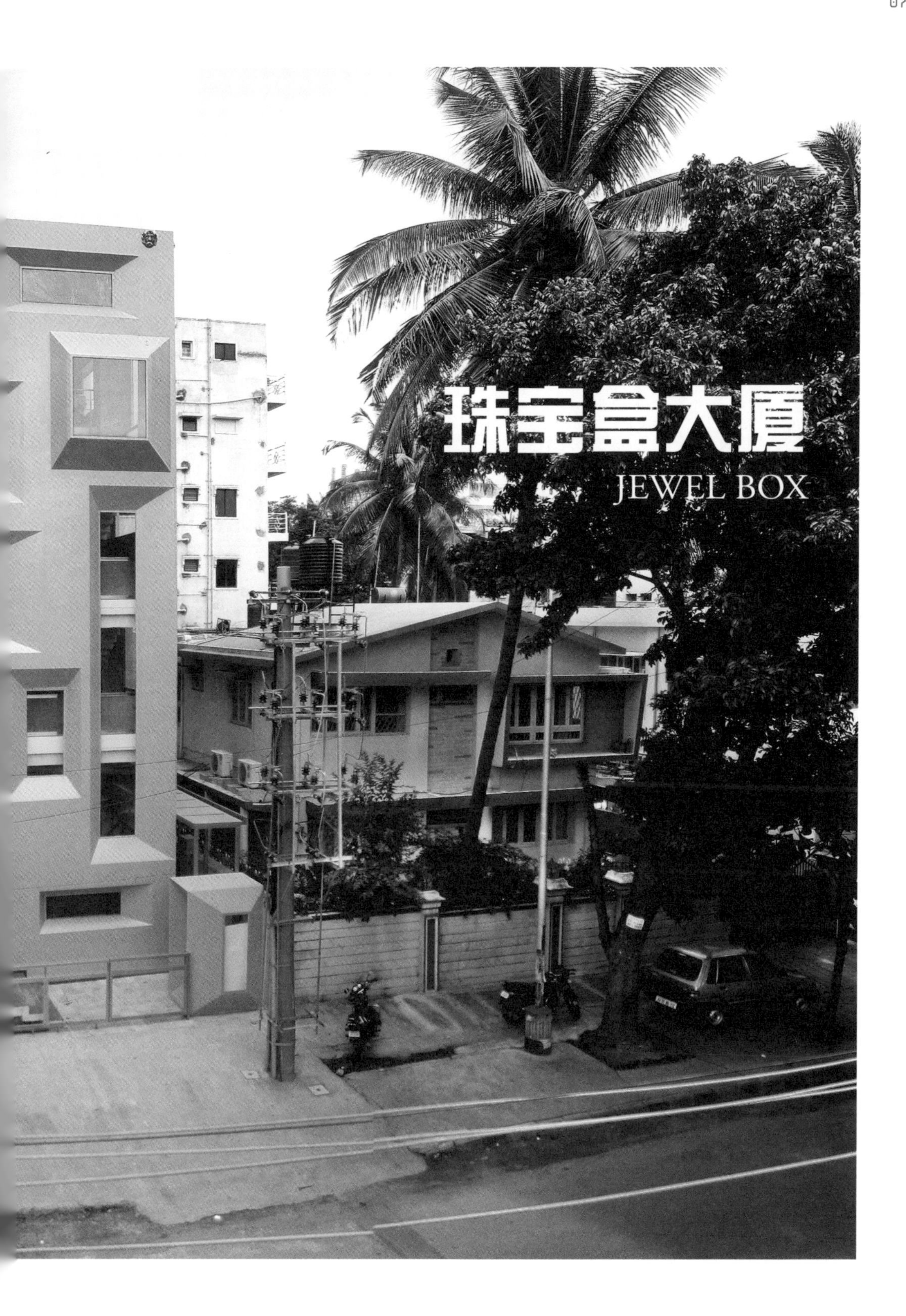

珠宝盒大厦
JEWEL BOX

项目名称：

珠宝盒大厦

Project Name:

JEWEL BOX

项目设计： SDeG
项目地点： 印度，班加罗尔
立面材料： 钢丝网水泥
竣工时间： 2010 年
摄影： George (SDeG)

Architects: SDeG
Location: Bangalore, India
Façade Material: ferrocement
Completion: 2010
Photographs: George (SDeG)

黄金和钻石谁人不爱？金条的质感 + 砖石的完美切面 = 璀璨金盔甲，打造出物质女的“超级战衣”。珠宝盒大厦即是为建筑秀场打造的一件“超级战衣”，每一处“关节”衔接构思巧妙，连贯而又曲折；每一个斜面之间的角度各不相同，构成极具视觉冲击的建筑立面。

Who doesn't like gold and diamonds? The texture of gold bar + the perfect facets of bricks and stones = a shining golden armor, providing a 'super battle suit' for material girls. Jewel Box is such a suit in the fashion show of architecture. All the 'joints' are exquisitely composed, creating a continuous and undulating image. The angles formed by the facets are different, making an eye-striking façade.

© 郑亚男（Nancy Zheng）

珠宝盒大楼是一座销售宝石的商务楼，内部还设有行政办公室。建筑的功能与水晶的独特几何造型完美地结合在一起。

为了刻画出立面的延展性，建筑师在设计上花费了不少心思——最初的灵感来源于天然水晶石及其凹凸不平的裂面。这些裂面经过合理的抽象演化出现在了建筑的表皮，使大小不一的窗体与其内部功能巧妙地结合在一起。从外部看来，这些“斜面”掩饰了这种联系，使建筑从外观上保持了均匀的整体性。

The Jewel Box is designed to retail precious stones, and accommodate administrative offices. The building is shaped by both program and the geometries of crystals.

A series of design processes then dictate the malleability of the façades - the architect's first inferences stemmed from raw crystal and its highly fractured attributes. The architect then aimed to rationalize the facets through several stages of refinement while ensuring a perceptible relationship between fenestration and the internal functions. However on the outside, the "bevels" aspire to camouflage this relationship and become many parts of a homogeneous whole.

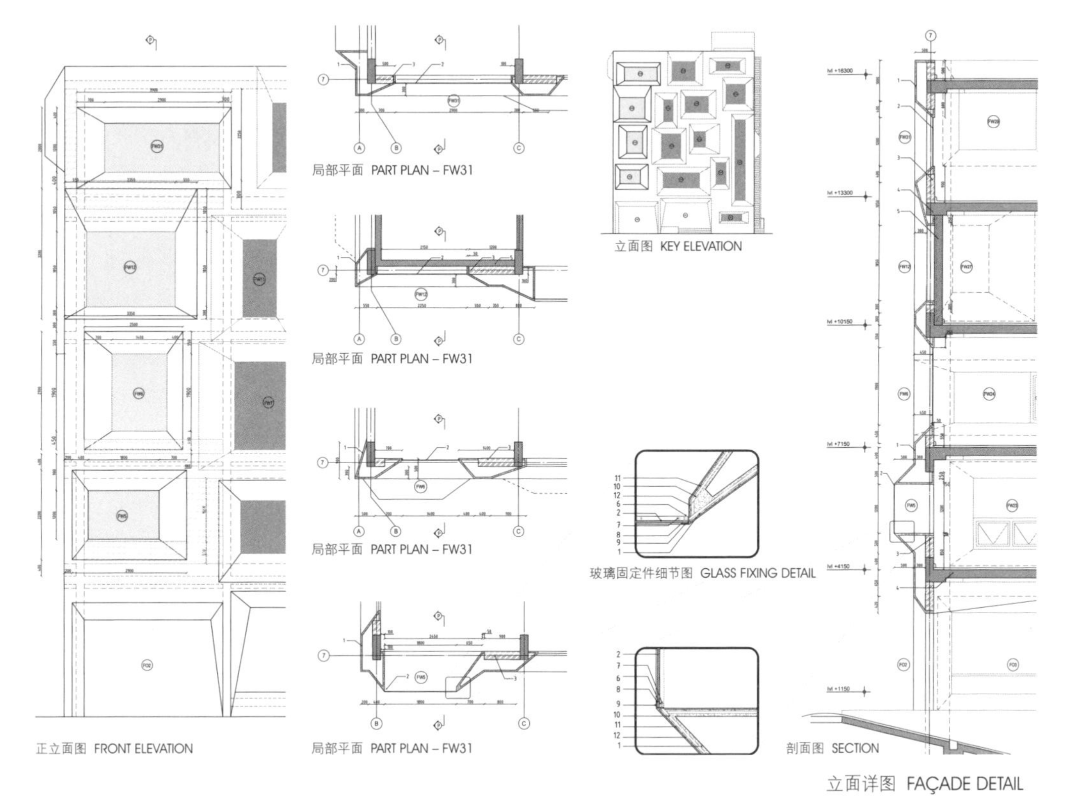

正立面图 FRONT ELEVATION

局部平面 PART PLAN – FW31

局部平面 PART PLAN – FW31

局部平面 PART PLAN – FW31

局部平面 PART PLAN – FW31

立面图 KEY ELEVATION

玻璃固定件细节图 GLASS FIXING DETAIL

剖面图 SECTION

立面详图 FAÇADE DETAIL

1 60 mm 钢筋混凝土
2 13.5 mm 夹层玻璃
3 200 mm 砌石
4 碾压混凝土板
5 200 mm 剪力墙
6 20 mm × 20 mm 铝珠
7 方头螺钉
8 3mm 厚双面胶
9 ISA 40 焊接于 Ø12 mm 混凝土杆
10 Ø12 mm 混凝土杆
11 Ø8 mm 混凝土杆
12 镀锌网

1 60 mm thick Ferroconcrete
2 13.5 mm thick laminated clear glass
3 200 mm thick masonry
4 RCC flat slab
5 200 mm thick shear wall
6 20 × 20 mm Aluminum beading
7 Coach screw
8 3mm thick double-sided tape
9 ISA 40 welded to 12 mmØ reinforcement rods
10 12 mmØ reinforcement rod
11 8 mmØ reinforcement rod
12 Galvanized chicken mesh

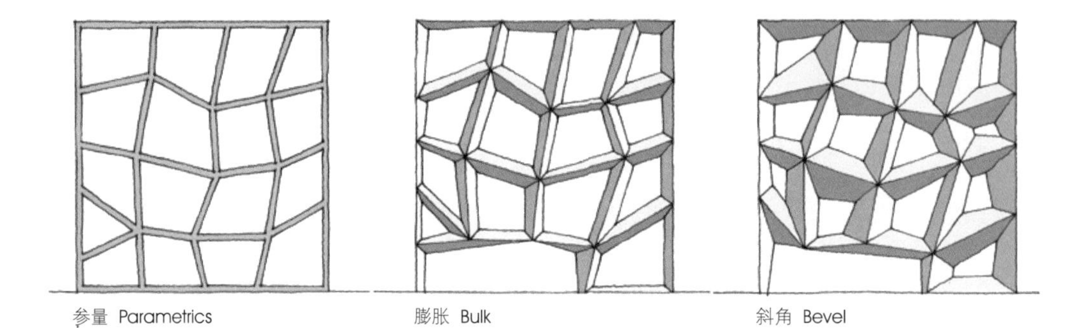

参量 Parametrics　膨胀 Bulk　斜角 Bevel

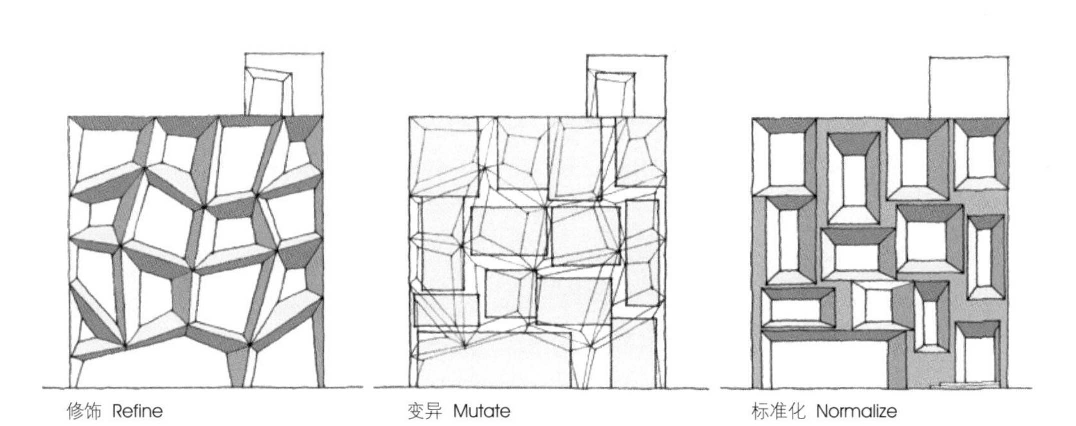

修饰 Refine　变异 Mutate　标准化 Normalize

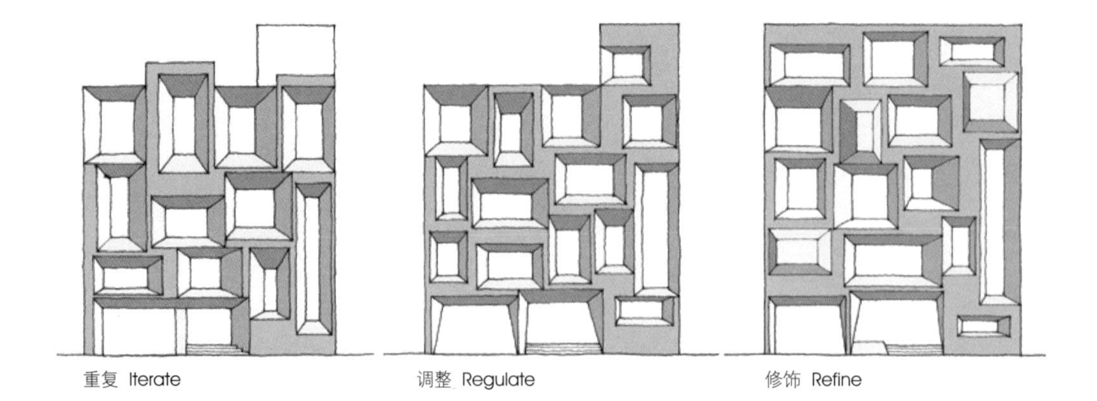

重复 Iterate　调整 Regulate　修饰 Refine

设计流程 SKETCH PROCESS

模型 MODELS

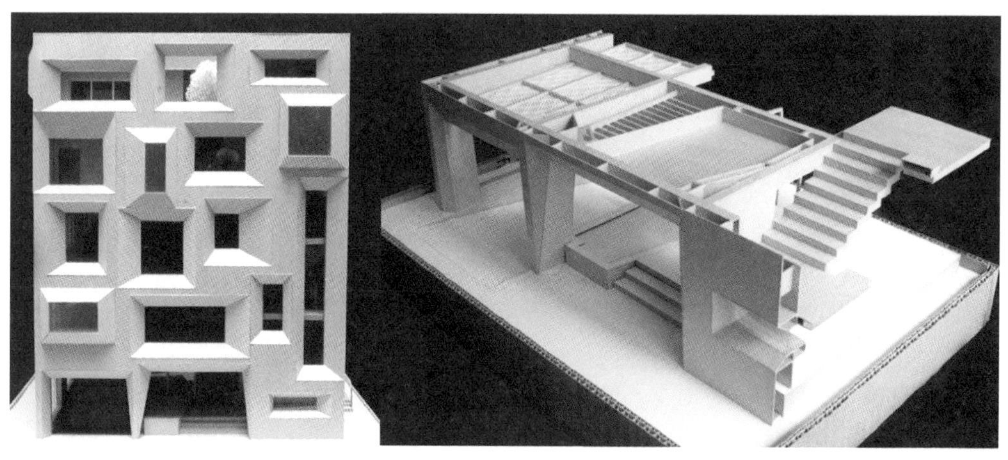

模型 MODELS

林林总总的小斜面构成 75 mm 厚的立面，采用钢丝网原地浇筑技术，轻盈地附着在钢筋混凝土主结构上。在建筑内部，这些"斜面"形成了丰富多样的窗棂框。因此，每个楼层拥有了多种视野，可供人们欣赏远近不同的窗外景色。

The beveled façade is 75mm thick, cast-in situ using ferrocement technology and light enough to be hung off the main RC structure. On the inside, the "bevels" create a variety of edge conditions, therefore, each floor therefore gets a different set of views - of the immediate and distant context.

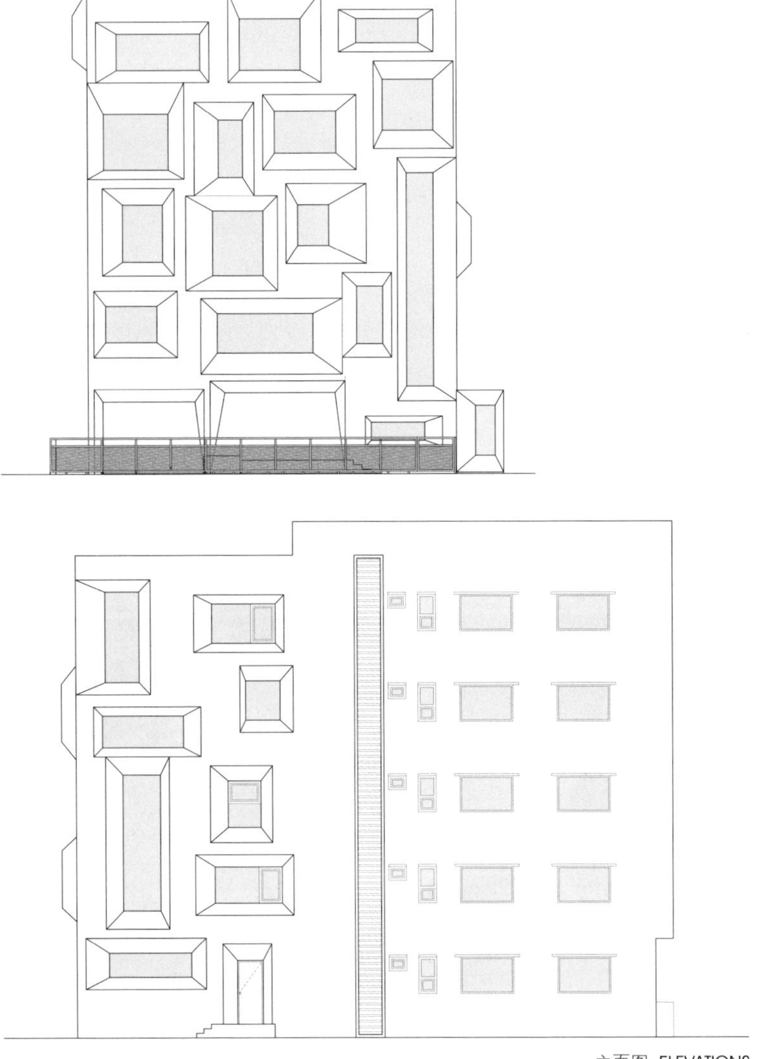

立面图 ELEVATIONS

塔斯马尼亚大学医学院与孟席斯研究所

SCHOOL OF MEDICINE AND MENZIES RESEARCH INSTITUTE, UNIVERSITY OF TASMANIA

项目名称：

塔斯马尼亚大学医学院与孟席斯研究所

Project Name:

SCHOOL OF MEDICINE AND MENZIES RESEARCH INSTITUTE, UNIVERSITY OF TASMANIA

项目设计： Lyons
项目地点： 澳大利亚，塔斯马尼亚
立面材料： 混凝土
摄影： Dianna Snape Photography

Architects: Lyons
Location: Tasmania, Australia
Façade Material: concrete
Photographs: Dianna Snape Photography

大自然，是设计师的灵感宝库。设计师从自然的山川河流中汲取创意灵感，将其丰富的造型、绚丽的色彩引入设计。服装设计中，取材于自然的图案与色彩往往经久不衰。线条蜿蜒曲折，韵律十足，温暖的土色与清爽的湖蓝色搭配，有回归自然之感。建筑设计中也追求与自然的和谐。这座大学建筑以混凝土装饰立面，流畅的拱形和倾斜的窗户象征着周围的山脉与河流，与周围景观融为一体。

Nature is like a treasure house for designers. They draw inspiration from mountains and rivers, and bring their various shapes and colors into design. For fashion design, the patterns and colors originated from nature always have enduring appeal. The serpentine lines contain a rhythm. The color of brown is as warm as the soil and blue as clear as a lake. The match of the two colors brings people back to nature. To be natural and harmonious is also what designers go after in architecture design. This university building has a stunning concrete façade. The fluid arches and slanted windows reference the mountains and rivers, allowing the building to blend into the surrounding landscape.

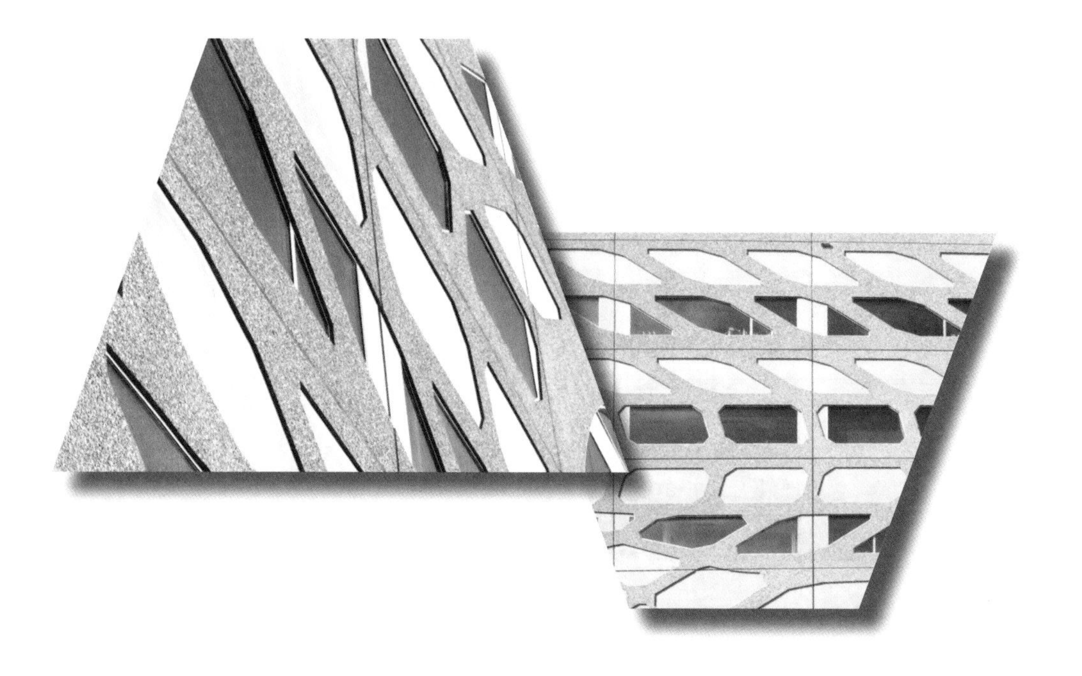

© 田静（Nicole Tian）

这座建筑拥有惊艳的混凝土立面，流畅的拱形和倾斜的窗户使立面特色鲜明。建筑的开窗布局源于对周围山脉和德文特河的抽象概括，将建筑意象表现得淋漓尽致。楼体的曲线造型参照了已不存在的溪流公园的设计，这座公园勾画了城市网络的边棱，并为新建筑提供了范例。

造型独特的窗户为上层住户提供了独特的视角，便于观赏壮美的景色。从街道望去，拱形的窗体参照了当地普遍采用的造型。同时，它那抽象的形体还象征着霍巴特城的山脉背景。

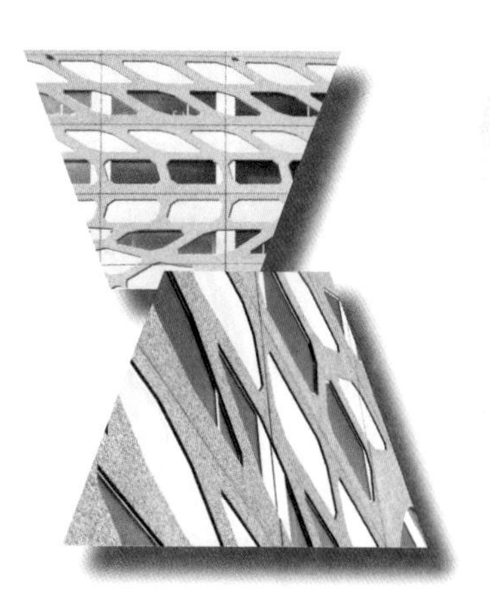

The building has an amazing concrete façade punctuated by flowing arches and slanted windows. The image of the building, expressed through its fenestration is derived, abstractly, from the surrounding mountain ranges and Derwent River. The curvilinear form of the building is a reference to the nonexistent Park Rivulet which was influential in shaping the edge of the city grid, upon which the new building is tied.

The shaped windows of the upper levels provide occupants with a means to see the spectacular landscape with new emphasis. On the street, the window "arch" forms reference an already established local typology whilst abstractly symbolizing the mountain ranges which background Hobart city.

立面图 ELEVATIONS

建筑以五星级绿色建筑评价标准为准绳，采取了诸多环保型策略，包括：尽量减少使用挥发性有机化合物材料；使用混凝土和可回收木地板；用镀锌钢代替铝材；安装变风量系统和通风柜；使用循环热回收系统、建筑自动化系统、太阳能热水系统、能源和水计量器、日光和人工照明控制系统以及高效的照明系统。

Green Star 5 Star was used as a reference rating tool to establish the extensive environmental strategies for the building. These included: minimized VOCS, concrete and recycled timber flooring, galvanized steel in lieu of aluminum, variable air volume systems and fume cupboards, run around coil heat recovery systems, building automation system, solar water system, extensive energy and water metering, daylight and artificial lighting control, high efficiency lighting.

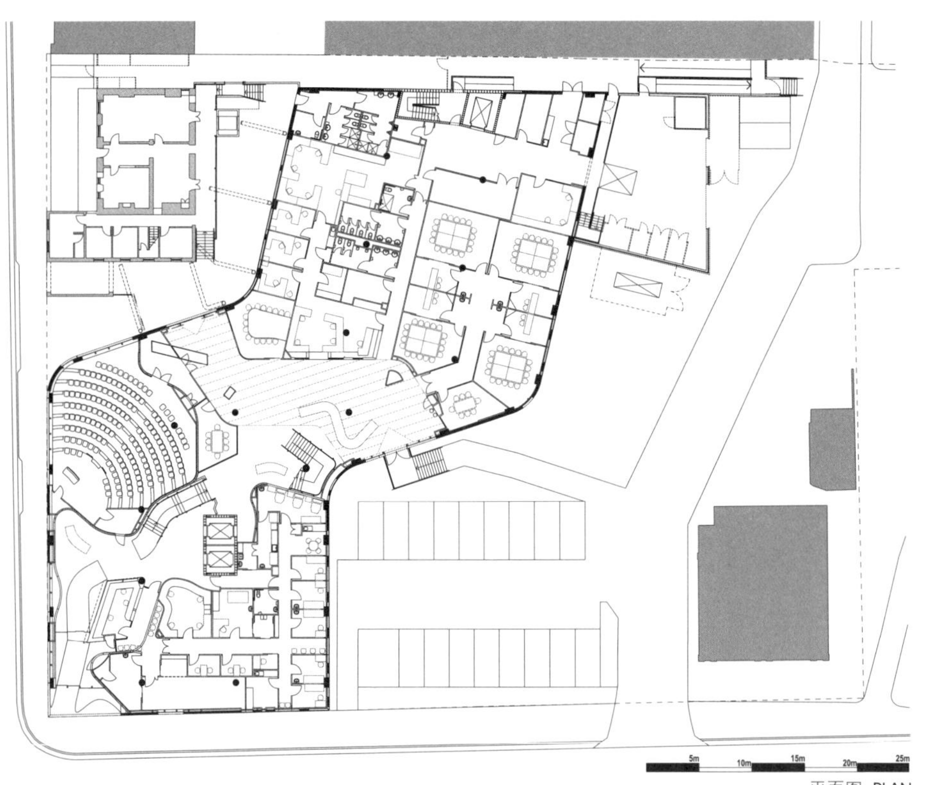

平面图 PLAN

HOLLYDE

UNIVERSITY OF TASMANIA
CITY
MEDICAL
SCIENCE
PRECINCT

褶皱墙
CRUSHED WALL

项目名称：

褶皱墙

Project Name:

CRUSHED WALL

项目设计： Walter Jack Studio
项目地点： 英国，康沃尔
立面材料： 混凝土
竣工时间： 2012 年
摄影： Simon Burt

Architects: Walter Jack Studio
Location: Cornwall, UK
Façade Material: concrete
Completion: 2012
Photographs: Simon Burt

华丽的衣服、名贵的首饰、精致的妆容，如此堆砌起来的矫作美人，看多了终会心生腻味，更加向往清新自然之风。摒弃庸俗，拒绝炫耀，回归含蓄简约风！用简单的设计、基本的色调打造随性慵懒之风，尽展曼妙柔美的身姿。"清水出芙蓉，天然去雕饰"，无浓妆艳抹之妖艳，亦无矫揉造作之华丽，却依旧魅惑天成。迷离的眼神、曼妙的身段、慵懒的姿态，无声地诉说着独属她的那份性感与迷人。

People will eventually get bored with the beauties piled up with resplendent clothes, expensive jewelries and exquisite makeup. They therefore yarn for something fresh and natural. They try to abandon vulgarity and coxcombry and return to a simple, reserved style. With simple design and basic colors, a casual style can also fully show the graceful figure. "Natural beauty of hibiscus is rising out of clear water." Though without the coquettish and luxuriance resulted from heavy makeup and artificiality, a sort of glamour is created naturally. Her unique appeal and charm are silently revealed through the sleepy eyes, graceful figure and relxed posture.

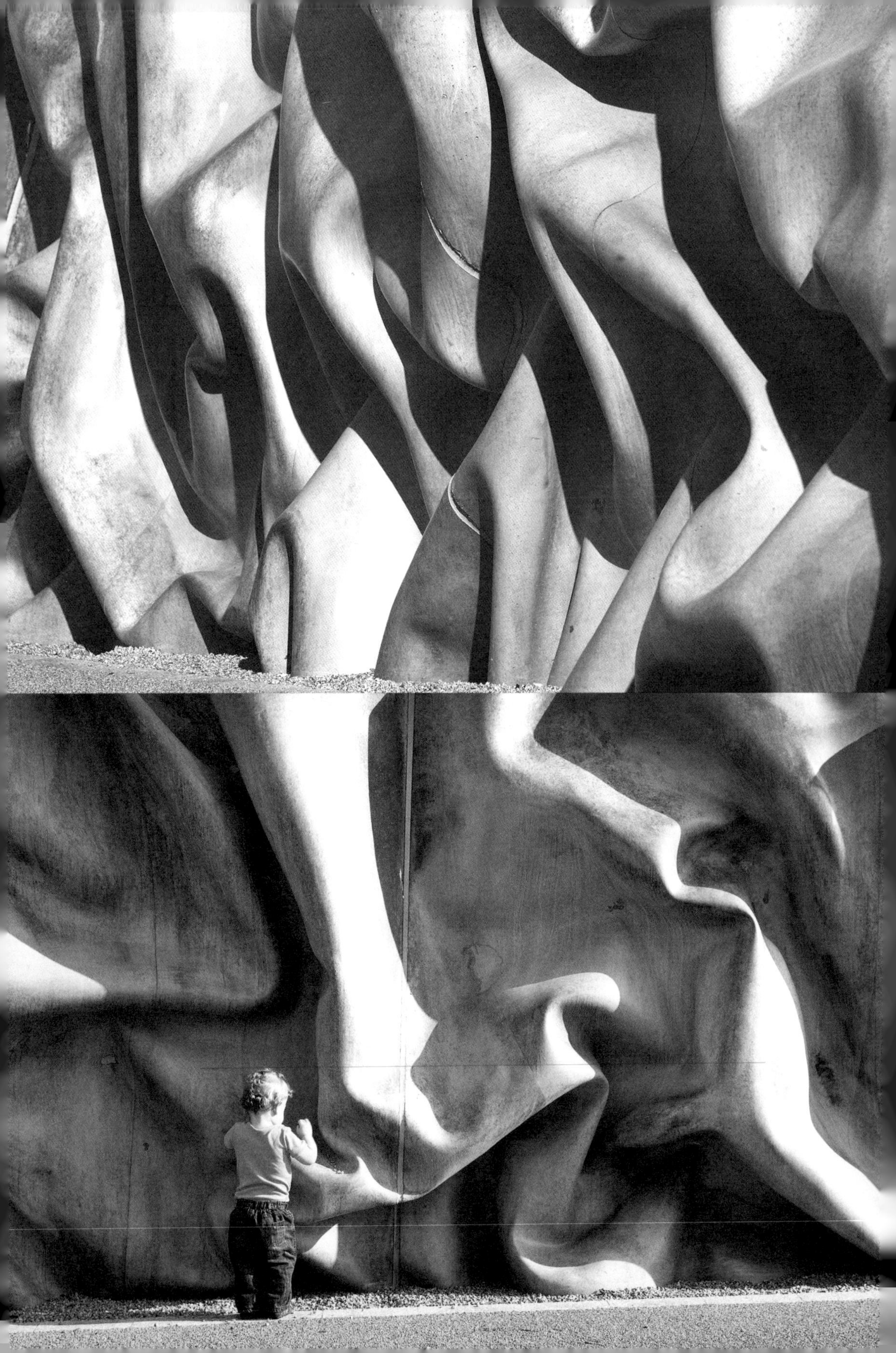

皱褶墙是一道长 30 m、高 3 m 的挡墙，它从 Pool 村一直延绵到该地区的中心地带，是该地矿业遗址的缩影。墙体一端高于另一端，象征着该地区经挤压形成的独特地貌。墙体是由一个长 30 m 的胶合板模架和一块橡胶长薄板构筑形成。首先将薄板进行褶皱和折曲，制成大型的胶状模具，以便将混凝土浇灌于其中。然后将模架切割成 6 个部分，并运往混凝土灌注入模的莱德混凝土厂。混凝土凝固后，将其从模具中取出，并运往不到 1.6 km 处的核心区进行安装。

Crushedwall is a 30m long, 3m high structure leading from Pool village to the main Heartlands site. The site's former mining heritage is reflected in the wall. Higher at one end than the other, the wall reflects the crushing aspects of geology. Process of creation involves 30 m of formwork from plywood and a long sheet of rubber, which was crumpled and folded to create a giant jelly mould in which the concrete was set. Formwork was then cut into six sections and transported to Ladd's Concrete where the concrete was poured into the moulds. Once concrete was set and had been removed from the moulds, it was transported less than 1.6 km up the road to the Heartlands site for installation.

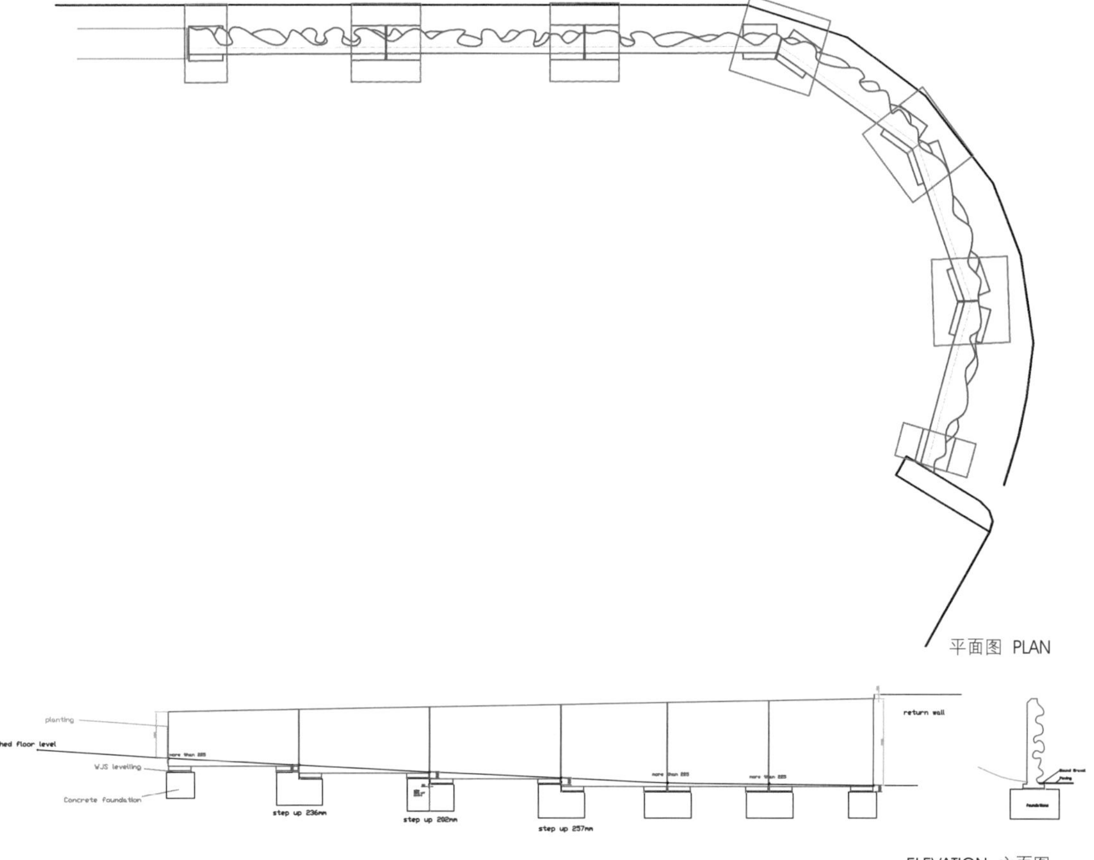

平面图 PLAN

ELEVATION 立面图

6个模具均为一次性产品，将它们切割开才能将成型的结构体取出。因此，每一个模具的混凝土浇灌工作必需一次成功。一旦浇灌出现失误就意味着要重新制作全部的6个模具。另外，还要对模具采取加固和支撑措施，以防止潮湿的混凝土产生流体静压而出现外凸。每一块结构体用了3~4个星期才浇筑成型，而安装工作一天之内就完成了。

混凝土并非以其流动性和柔软性著称。然而，此时混凝土的确是呈"液态"。设计者借此设计显示出混凝土的本质，保持了其流动性。

The six moulds were single-use - to get the structure out they had to be cut away - so there was a need to get each pour right first time, just one incorrect piece would require all six moulds to be remade. Extra precautions were taken to reinforce and support the moulds due to the hydrostatic pressure created by the wet concrete trying to push out. Each piece took three to four weeks to cast and were installed in one day

Concrete is not noted for its fluid softness. And yet it is a liquid. The designers wanted the concrete to tell its own story - to retain the liquidness of its process.

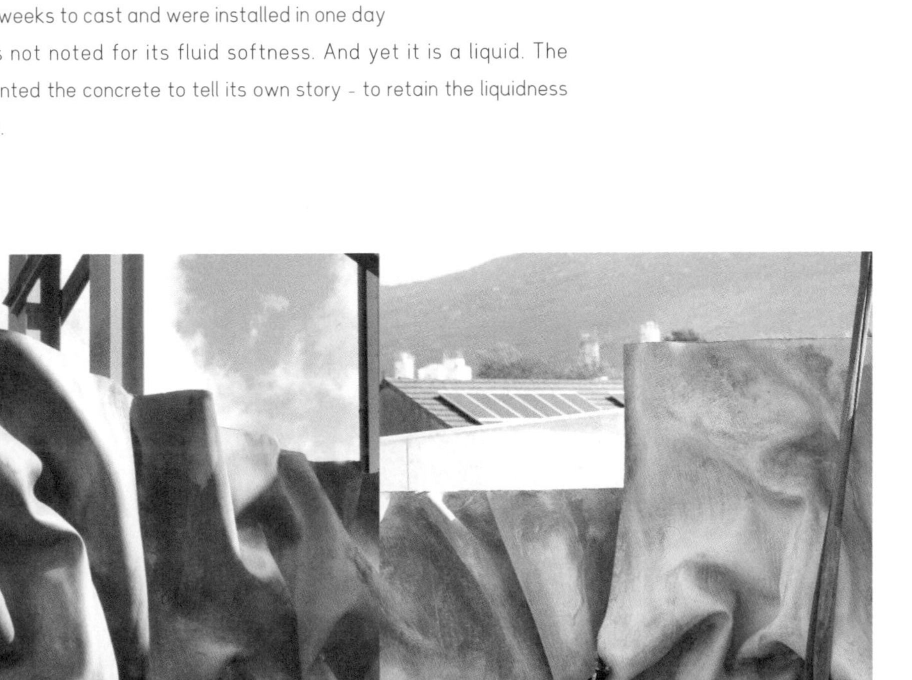

O-14 大厦

O-14 TOWER

项目名称：
O–14 大厦
Project Name:
O-14 TOWER

项目设计： RUR Architecture - Reiser + Umemoto
项目地点： 阿联酋，迪拜
立面材料： 混凝土
竣工时间： 2010 年
摄影： RUR Architecture

Architects: RUR Architecture - Reiser + Umemoto
Location: Dubai, United Arab Emirates
Façade Material: concrete
Completion: 2010
Photographs: RUR Architecture

从 20 世纪 50 年代开始风靡的波点，已成为经久不衰的时尚元素。波点图案看似千篇一律，实则千变万化，大波点象征着自由和活力，小波点则代表了浪漫和优雅，大小不一的波点错落分布，充满青春的俏皮与活泼。波点复古潮，长裙更优雅，一袭白色蕾丝长裙点缀镂空黑色波点，朦朦胧胧，尽显女人的优雅与妩媚。香肩裸露，又增添了一种艺术般的性感。建筑若以波点做装饰，必定魅力无限，这座 O–14 大厦覆盖着镂空混凝土外壳，形成了蕾丝般的效果，散发着阵阵复古的性感。

Dots have been all the rage since the 1950s and become an enduring fashionable element. Dots seem the same, but have infinite variations. Big dots represent freedom and vigor while the small ones symbolize romance and elegance. The dots of various sizes scatter randomly, brings out a sense of curtness and vivacity. The retro dots are attached to long elegant dress, making a white lace dress embellished with black hallow-out holes. Hazily, women's grace and glamour are fully presented. Moreover, the exposed shoulders add a sense of artistic sexuality. A building decorated with dots is undoubtedly stunning. O-14 Tower is covered with a concrete shell with hallow-out patterns, creating a lace-like effect, thus a type of 'retro appeal' spreads out everywhere.

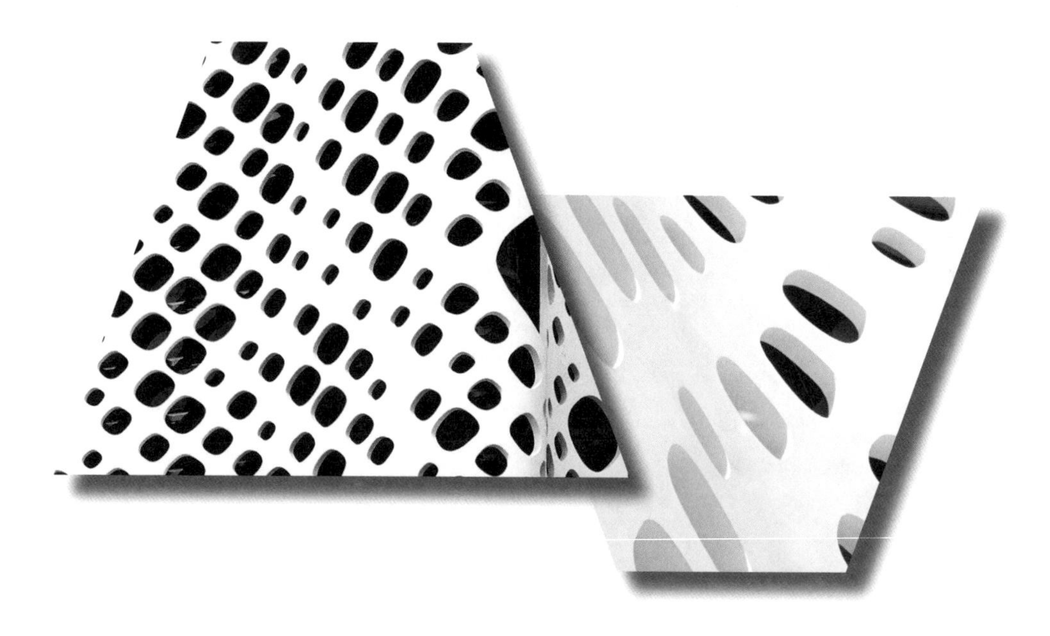

© 郑亚男（Nancy Zheng）

0–14 覆盖着 40 cm 厚的混凝土外壳。外壳上开有 1300 个孔洞，在建筑立面上形成了网眼蕾丝般的效果。制作这些孔洞首先要将计算机数字切割的孔状聚苯乙烯板导入钢筋模具中，并在浇筑混凝土之前在两侧装上钢质滑动模板。然后将超级混凝土浇筑在这个坚固的网格结构中，从而制成了精美雅致的网眼状外壳。

大厦的混凝土外壳提供了一个有效的结构外壳，能够让建筑核心免受侧力影响，并为建筑内部创造了无柱的宽阔空间。这个外壳不仅是建筑的结构要素，它还是一个透光、透气、视野通透的遮挡板。外壳上的这些孔洞是根据结构、视线、日照和亮度的需要而设置的。

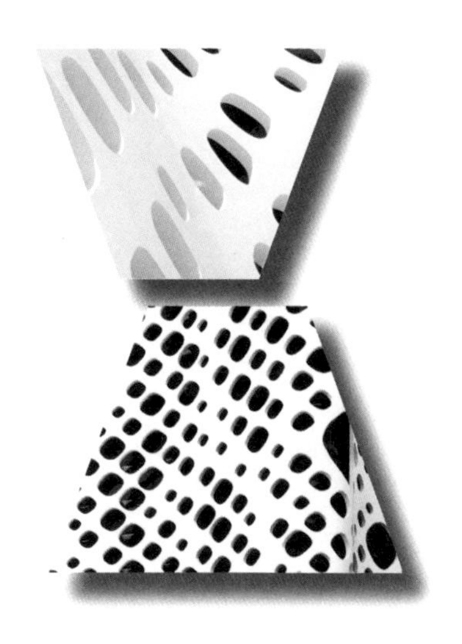

O-14 is sheathed in a forty centimeter-thick concrete shell perforated by over 1,300 openings that create a lace-like effect on the building's façade. The holes are achieved by introducing computer numerically cut polystyrene void forms into the rebar matrix, and sided with modular steel slip forms prior to the concrete pour. Super-liquid concrete is then cast around this fine meshwork of reinforcement and void forms resulting in an elegant perforated exterior shell.

The concrete shell of O-14 provides an efficient structural exoskeleton that frees the core from the burden of lateral forces and creates highly efficient, column-free open spaces in the building's interior. The shell is not only the structure of the building, it acts as a sunscreen open to light, air, and views. The openings on the shell modulate depending on structural requirements, views, sun exposure, and luminosity.

整体图案并非根据固定的模式而设定的（具有可变性）。它虚实相间，影响到了楼层具体功能的设置。外壳和建筑主体之间近 1 m 宽的空隙形成了所谓的"烟囱效应"，即热空气拥有向上的疏导空间，从而使网眼状外壳后的玻璃窗表面得以冷却。这种被动式太阳能技术是 O-14 大厦冷却系统中一种重要的自然冷却方式，它减少了建筑的能耗和成本。这只是建筑设计中众多创新性措施之一。

The overall pattern is not in response to a fixed program, (which in the tower typology is inherently variable), rather the pattern in its modulation of solid and void will affect the arrangement of whatever program comes to occupy the floor plates. A space nearly one meter deep between the shell and the main enclosure creates a so-called 'chimney effect', a phenomenon whereby hot air has room to rise and effectively cools the surface of the glass windows behind the perforated shell. This passive solar technique essentially contributes to a natural component to the cooling system for O-14, thus reducing energy consumption and costs, just one of many innovative aspects of the building's design.

拼贴展馆
PATCHWORK PAVILION

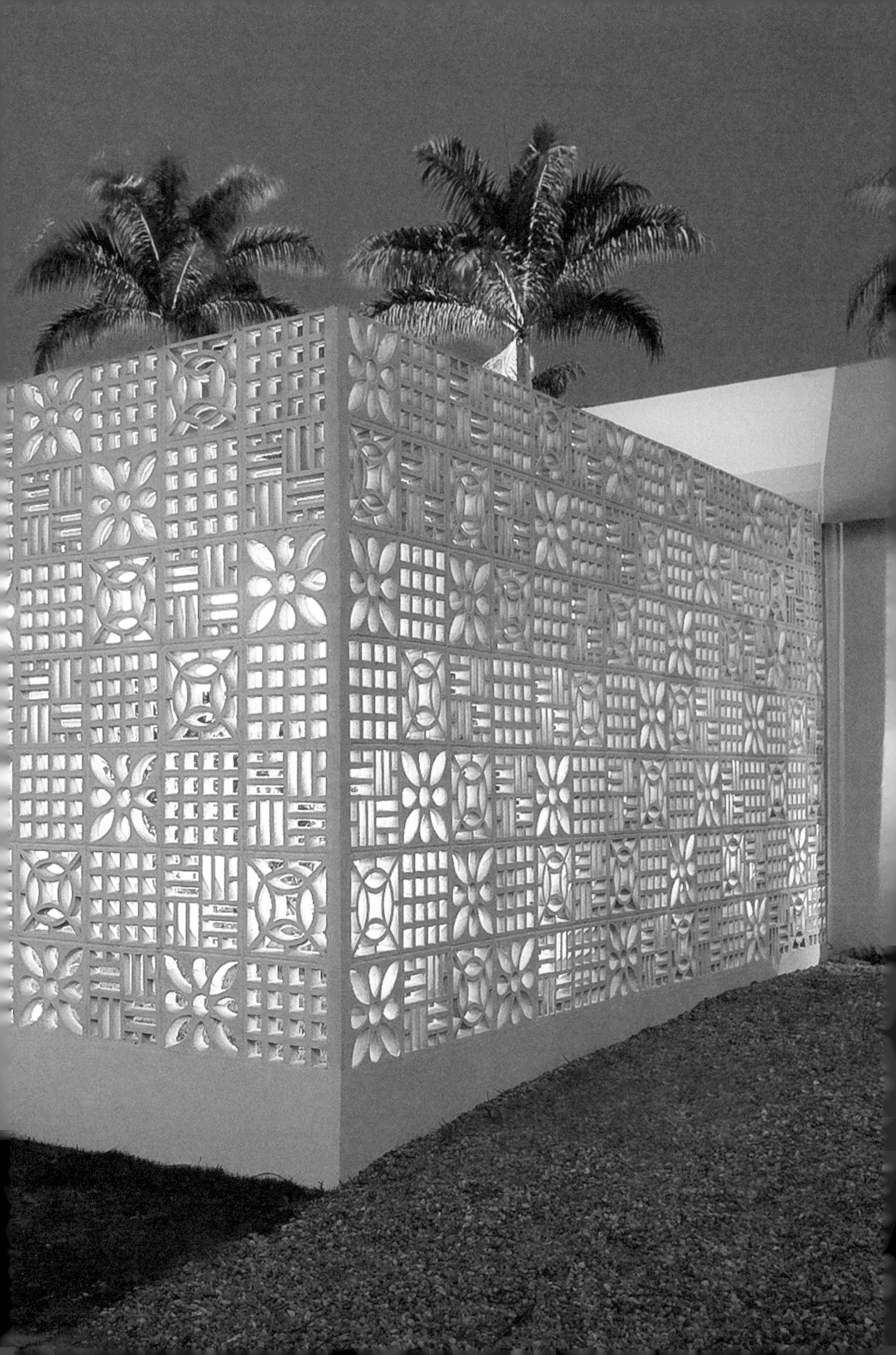

项目名称：
拼贴展馆
Project Name:
PATCHWORK PAVILION

项目设计： DOMO arquitetos
项目地点： 巴西，巴西利亚
立面材料： 预制混凝土砖
摄影： Frank Carvalho

Architects: DOMO arquitetos
Location: Brasilia, Brazil
Façade Material: precast concrete blocks
Photographs: Frank Carvalho

用充满设计感的金色和黑色相互交织而成别致的镂空图案，与中式旗袍元素相结合，这件上衣给人以华丽富贵之感，把女性的优雅柔美表现得淋漓尽致。这座拼贴展馆俨然是一位优雅的贵妇人，白天，原本带有民族风格图纹的白色镂空墙面显得如此宁静安然；到了夜晚，金黄色的光晕穿过镂空蔓延开来，逐渐隐秘到夜色当中，使得"贵妇人"又增添了几分高雅与神秘。

The exquisite sleeveless shirt is specialized by its hallow-out patterns interweaving golden wit h black colors. Integrated with Chinese cheongsam elements, this shirt gives a gorgeous and prestigious feeling, and fully shows female's grace and delicacy. This pavilion is just like a graceful lady. In the daytime, the white hallow-out walls with ethnic patterns look so quiet. When night falls, the golden light spreads out through the hallow-out walls, adding more elegance and mystery to the pavilion.

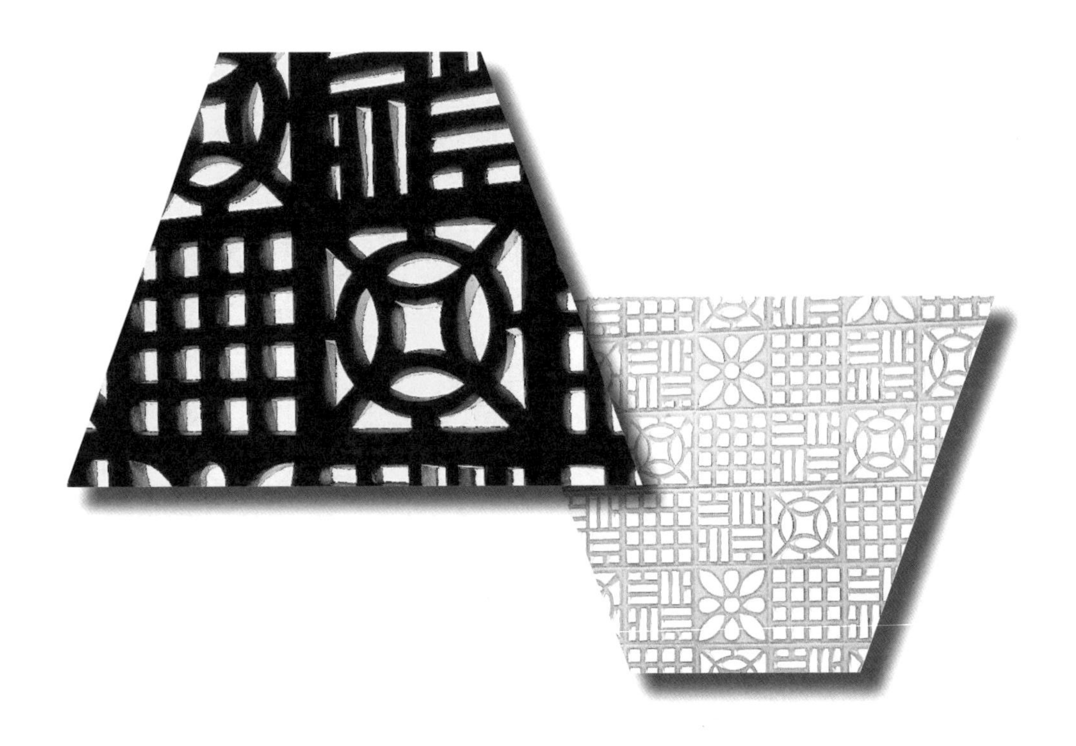

“概念创意”
"Concept"

© 宋宁（Bella Song）

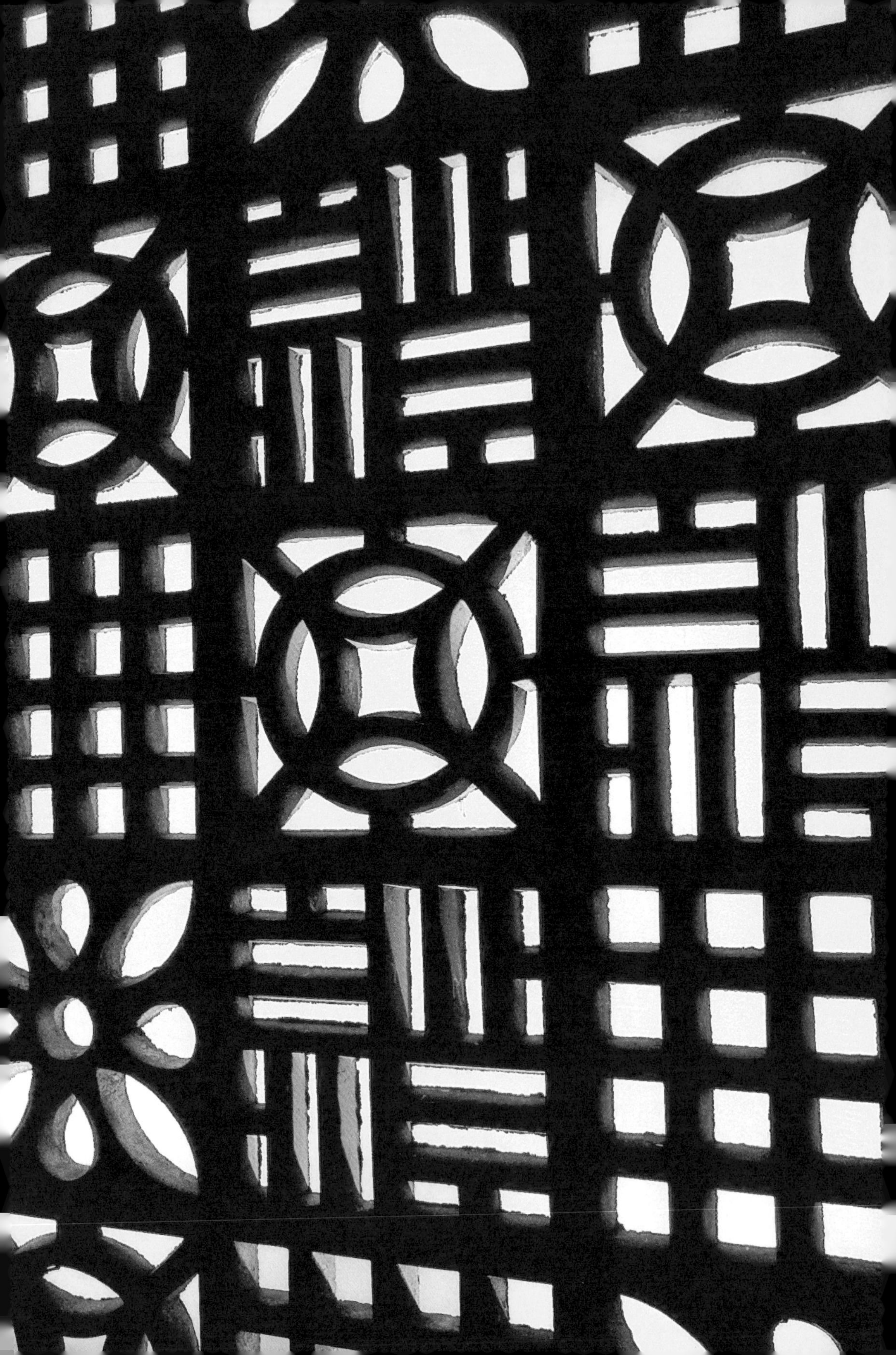

该设计旨在打造一个朦胧的、半透明的绘画展览空间，并在旁侧建成一个无顶的花园用来展览雕塑作品。刻有各种图案的预制混凝土板随意地搭砌在一起，形成一层半透明的外壳，包裹整个建筑体。图案参照了巴西东北部地区工匠所创制的拼缀刺绣物“Rendado”。展厅的外墙用预制混凝土砖构筑，内部墙壁由钢架、钢化玻璃和石灰构成，画廊的地面则用天然巴西石材铺砌。

展馆的主入口是一条坡道，它清晰地将展馆不透明的黑色部分和半透明的白色部分分隔开来。黑色的盒状建筑拥有双层墙壁，内部设有黄色的内衬。夜晚，灯光照亮了这个黄色的空间，将这道墙壁变成了展馆其他部分的方位灯。墙壁上的小孔能够让游人在穿过入口走廊时瞥见展馆内部的情形。

The basic idea was to create an exhibition space that was half opaque and half transparent for exhibiting the paintings and a roofless garden by its side for sculptures, all wrapped by a "translucent skin" made of precast concrete blocks of different patterns placed in a random combination. The resulting overall pattern makes reference to the patchwork embroidered pieces made by artisans from the northeastern part of Brazil, known as "Rendado". Precast concrete blocks of different patterns were employed for the outer wall, steel frame, tempered glass, plaster walls and the natural Brazilian stone for the inner gallery flooring.

The main access to the gallery is made through a ramp that articulates the opaque black-colored part of the gallery with its white-translucent half. The black box is built as a double layered wall that contains a yellow colored surface inside. This yellow colored void is lit by indirect light during the night, transforming the wall into a orientation lamp for the rest of the exhibition. A small hole in this wall lets the visitor take a look at what happens inside it while he walks through the access corridor.

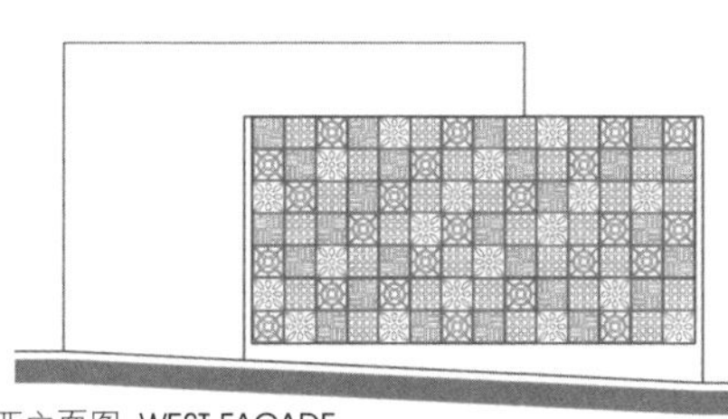

西立面图 WEST FAÇADE

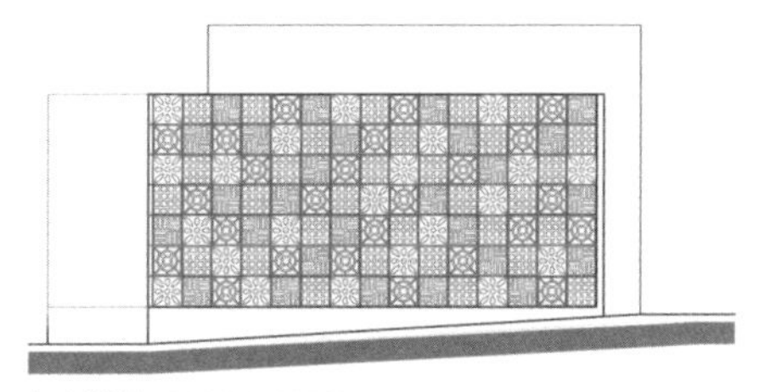

东立面图 EAST FAÇADE

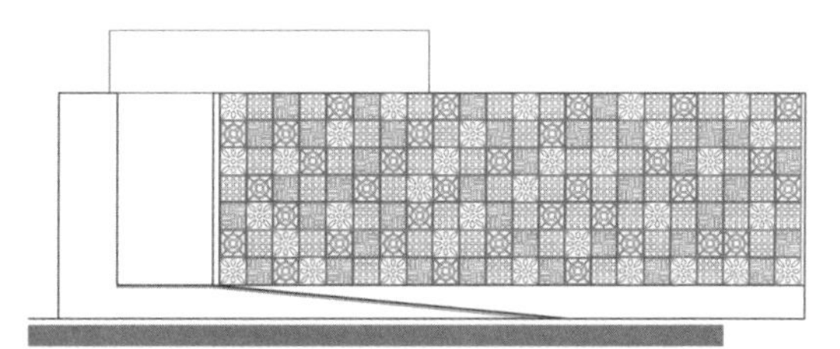

南立面图 SOUTH FAÇADE

因为便于自然通风和光照，所以预制混凝土板曾在 20 世纪 60 年代被广泛地用于巴西利亚早期的现代主义建筑中。在巴西，这也是一种造价低廉的材料。通过使用预制混凝土板——这种在极为巴西普遍的材料，建筑师创造出了一个半透明的、纹理繁复的临时展馆，而且成本低、用时短。

Precast concrete blocks were also largely used in the early modernist buildings of Brasília during the 60's, for allowing natural ventilation and light into the building. It is also a very low-cost material in Brazil. Using pre-cast concrete blocks-ubiquitous in Brazil-the architects created a temporal structure that was half opaque and half translucent, rich in texture and sophistication, but done affordably and on schedule.

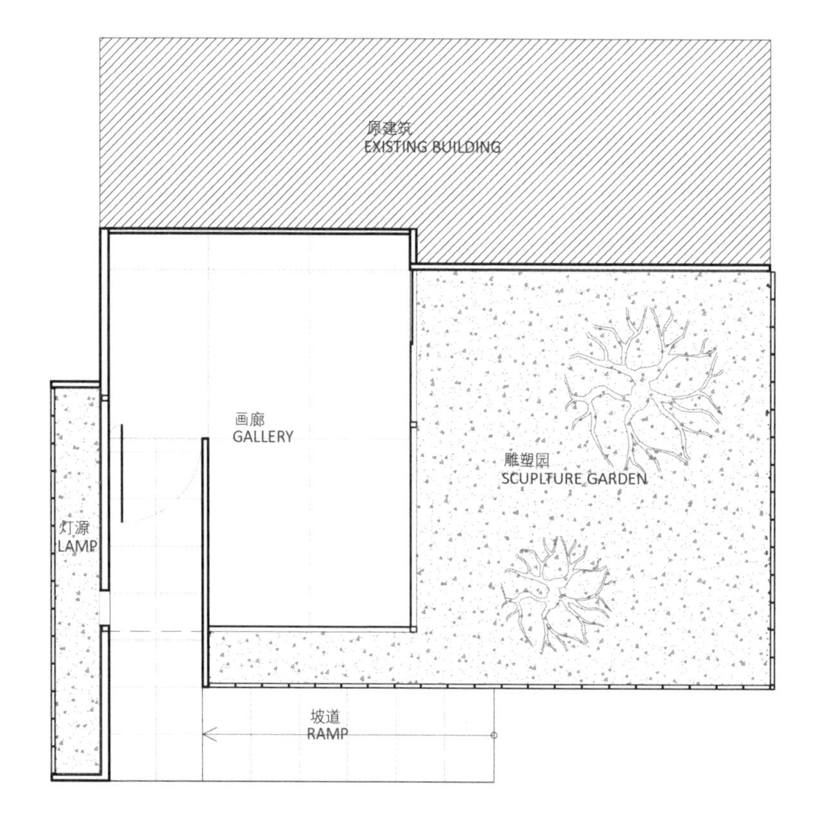

一层平面图 GROUND FLOOR PLAN

皮瑞伯德公寓楼

PREBOLD APARTMENTS

项目名称：

皮瑞伯德公寓楼

Project Name:

PREBOLD APARTMENTS

项目设计： Studio Kalamar
项目地点： 斯洛文尼亚，皮瑞伯德
立面材料： 纤维水泥板
摄影： Miran Kambič

Architects: Studio Kalamar
Location: Prebold, Slovenia
Façade Material: fiber cement
Photographs: Miran Kambič

丹麦哥本哈根湛蓝的海边，小美人鱼坐在岩石上，面向大海，述说着一段纯净、执著而伤感的爱情。她身着鱼鳞片的抹胸裙装，搭配长长的鱼尾裙摆，在月光下泛着银色的光。她用深沉而忧凄的目光望着苍茫的大海，似在等待着爱人的归来，纵然潮起潮涌，她默默守候，朝朝暮暮。

The little mermaid sits on a rock along the blue seashore in Copenhagen, Denmark. Facing the sea, she is telling a pure, dedicated but sad love story. She wears a strapless dress made of scales, with a long fishtail hemline, shining silver light in the moonlight. She stares at the boundless sea with deep, sad eyes as if she is waiting for her lover's return. She waits there quietly day and night with the ebbs and flows of the sea.

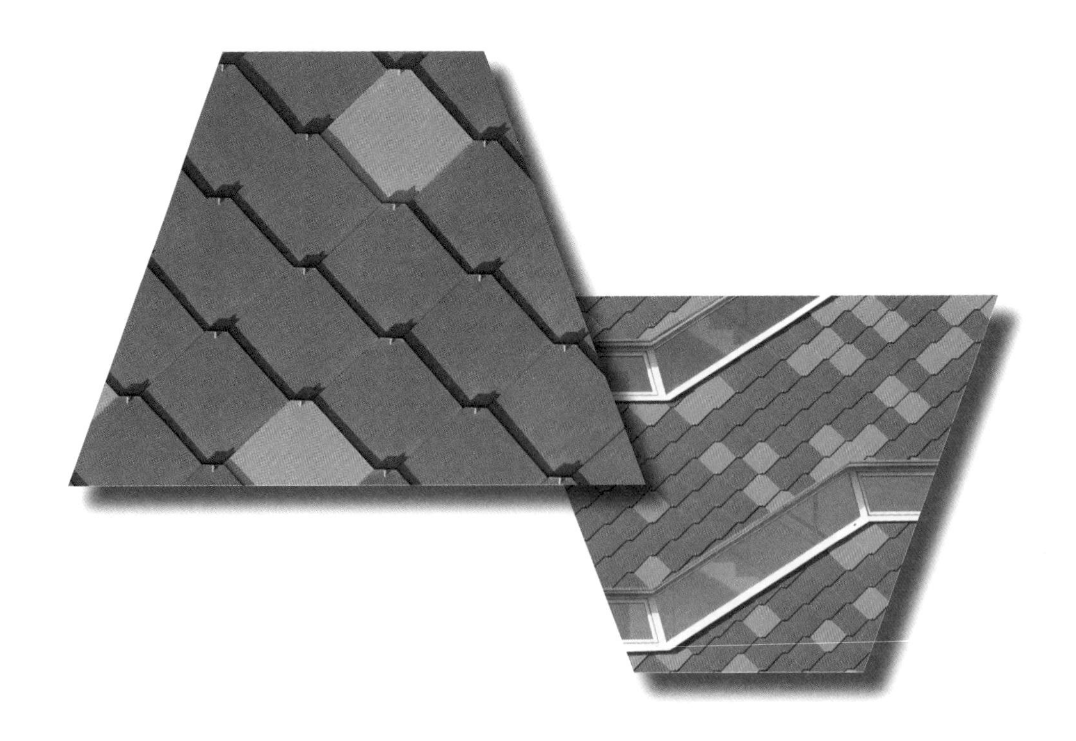

"概念创意"
"Concept"

© 宋宁（Bella Song）

这座公寓楼坐落在一个乡村小镇的边缘地带。大楼西边朝向一片种植着啤酒花的广袤田野。蜱酒花是整座山谷的特色农产品。

鉴于建筑的立面较为平直，设计者从这座山谷的代表性农产品啤蜱酒花中撷取了灵感，选用了鳞片结构的表皮，一直铺展至屋顶。它由两种颜色的纤维水泥板拼接而成，与土地的颜色形成了呼应。

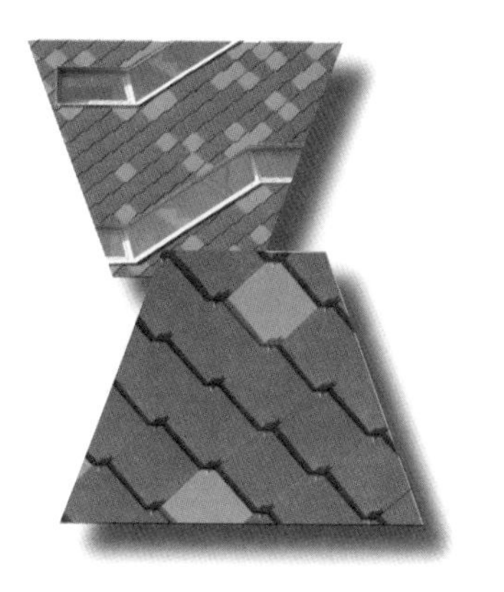

The apartment building is located on the edge of a small town in a rural environment. To the west, the building is oriented to large hops fields, hops being the specific agricultural product of the entire valley.

In view of planar façade surfaces, a scaly structure has been chosen, derived from the form of the typical valley produce, the hops cob. Scaly structure of the building is materialized in material, purposed for covering roofs - sheets of fiber cement in two colours, echoing surrounding earth hues.

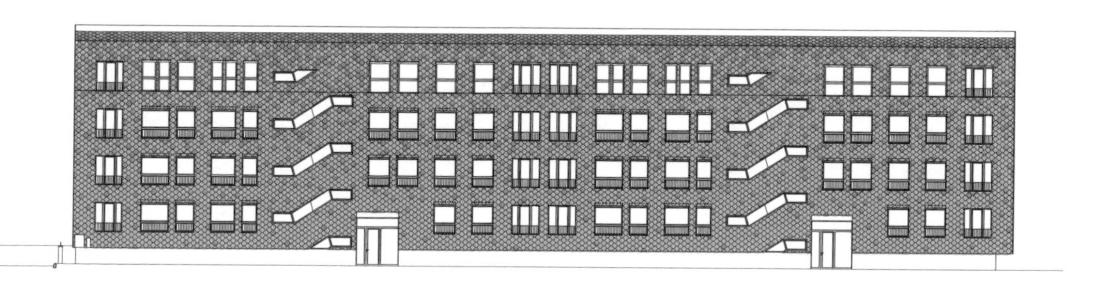

正立面图 FRONT ELEVATION

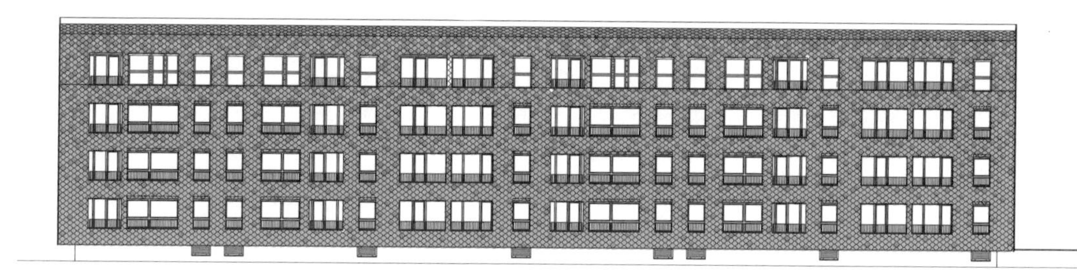

后立面图 BACK ELEVATION

场地的位置决定了建筑需采用南北走向，同时也保证了良好的采光条件和宽阔的视野。入口和停车场均位于场地的东侧，建筑狭长的西立面则朝向蜱酒花种植地。建筑的顶部向上收拢，颇具动感，这使得建筑方正的体量显得十分轻盈。这种建筑表现形式更适合现在的城市尺度。

Site position determines volume's north-south orientation which also ensures quality light conditions and good views. Access and an-site parking are both arranged on the eastern site, the long western elevation is oriented to hops fields. A dynamic apex of the volume results in a lighter and cubic architectural expression more appropriate to the existing urban scale.

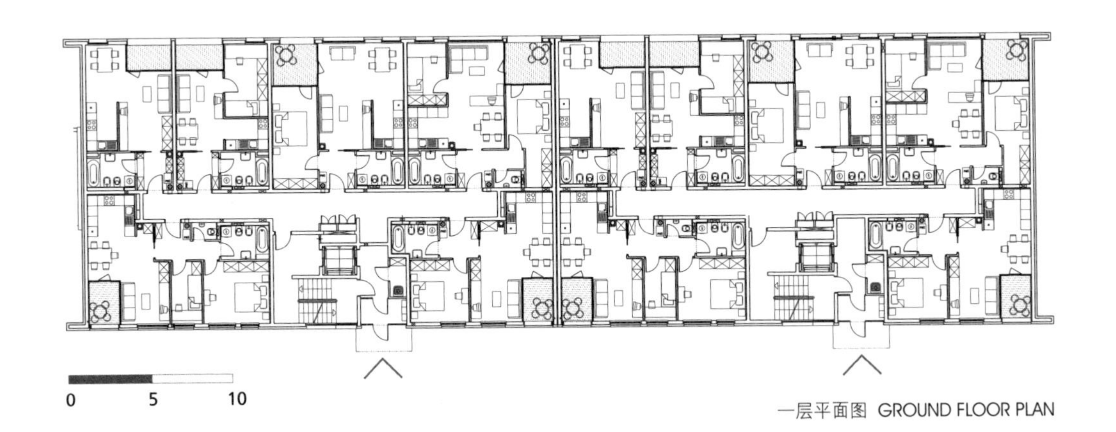

一层平面图 GROUND FLOOR PLAN

100K 住宅 100K HOUSE

项目名称：

100K 住宅

Project Name:

100K HOUSE

项目设计： Interface Studio Architects
项目地点： 美国，费城
立面材料： 纤维水泥板
竣工时间： 2010 年
摄影： Interface Studio Architects 友情提供

Architects: Interface Studio Architects
Location: Philadelphia, USA
Façade Material: fiber cement
Completion: 2010
Photographs: Courtesy of Interface Studio Architects

蓝色代表忧郁，白色象征纯洁，两种色彩搭配起来就像是蓝天和白云，给人以清新脱俗的感觉。蓝白相间的长裙优雅飘逸，宽大裙摆隐隐闪着银光，如星星般璀璨耀眼，裙身的线条如流水泻落，勾勒出窈窕身姿。阳光下翩翩起舞，动静中流光溢彩。100K 住宅仿佛穿上了这样一件蓝白相间的华服，设计师运用简单的材料和图案，打造出流光溢彩的立面。

The color blue represents the feeling of melancholy and white stands for purity. The combination of these two colors is like the sky and the clouds, giving people a fresh feeling. The long dress in blue and white is graceful and elegant, with the broad hemline glittering with silver light, sparkling like stars. The curve of the dress is as fluent as water, perfectly articulating the graceful figure. When dancing in the sun, it seems light and colors are flowing around. 100K House looks like in such a fancy blue-white dress. The architect creates a fascinating fa ade with simple materials and patterns.

© 施梅玲（Meiling Shi）

在 100K 住宅项目中，建筑师着力打造紧凑、隔热性能强的建筑外壳，只需安装小型的暖通空调系统，从而降低了能耗。建筑师对图案、纹理和色彩进行了实验，最终在 100K 住宅的立面上采用了极为平整的纤维水泥覆层。

In the 100K Houses, the architects focused on creating a tight, well-insulated building envelope, which requires a smaller HVAC system and lowers energy costs. Experimenting with pattern, texture, and color, they employed super flat material variations in fiber cement cladding on the 100K House façades.

超性能外壳
100K 住宅的高性能品质之一在于其高效保温、密封的外壳。下面的剖面图代表了两种不同的构建方法。

SUPER TIGHT ENVELOPE
The high performance qualities of the 100K Houses start with a well-insulated, tightly sealed envelope. The sections below represent the two different construction methods we have used.

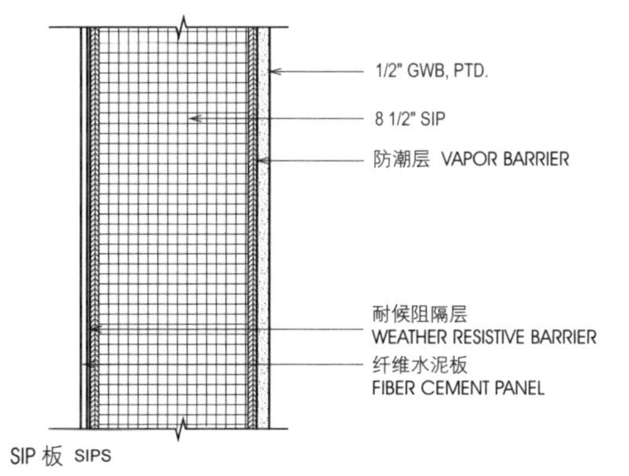

SIP 板 SIPS

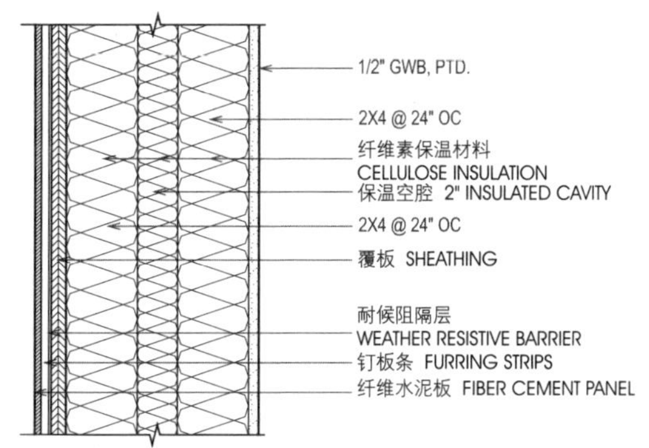

双面框架 DOUBLE STUD STICK FRAME

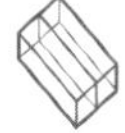
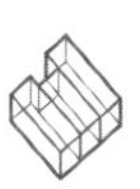

小巧。规模过大的住宅往往会采用廉价的材质和拙劣的设计，其经济效益也难以实现。小巧才是住宅设计的新标准。

简洁与平整。建筑师致力于建设高效、低成本的建筑，并且根据建筑场地位置因地制宜提出了设计方案。100K 住宅以简单的材料构筑了流光溢彩的立面，其纹理、图案和色彩等装饰虽成本低廉，却极具冲击力。

绿色。100K 住宅采用了被动式节能策略，着力于建筑表皮的性能，而非建筑系统。这一策略使建筑荣获 LEED 认证的住宅建筑铂金奖。其住宅能耗率标准在 42 和 24 之间，比按低能耗标准设计的住宅高出 58%~76%。

SMALL. Oversized homes are largely the result of cheap materials, poor-quality design, and unrealistic economic expectations. Small is the new normal.

SIMPLE AND FLAT. The architects define efficient, cost-effective construction and exploit that position for design opportunities. The 100K Houses riff on simple materials and flush façades, employing texture, pattern and color as low cost, high impact treatments.

SUPER GREEN. The 100K Houses employ a passive strategy focused on envelope rather than systems. This strategy yielded LEED for Homes Platinum certification and HERS ratings between 42 and 24 (performing 58% to 76% better than homes designed to baseline energy code).

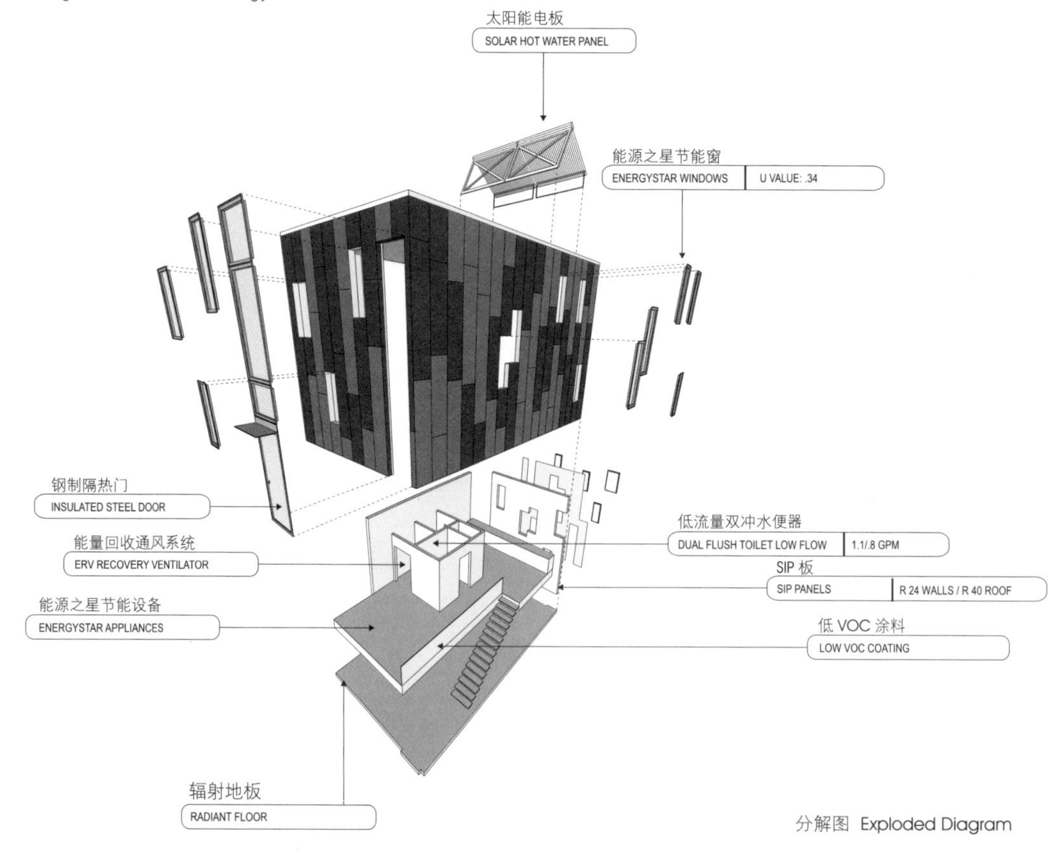

分解图 Exploded Diagram

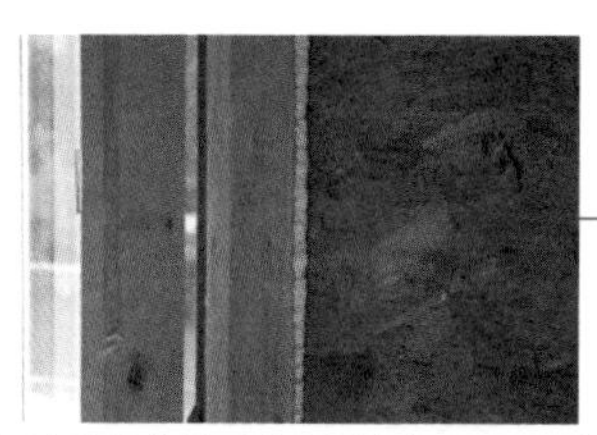

双夹墙组件显示（左到右）：2X4，空气层，2X4，喷涂泡沫

DOUBLE STUD WALL ASSEMBLY SHOWING (LEFT TO RIGHT): 2X4, AIR SPACE, 2X4, SPRAY FOAM

窗口大小与框架协调，以减少材料及施工浪费

WINDOW SIZES WERE COORDINATED WITH FRAMING TO REDUCE MATERIAL AND FIELD WASTE

Skinny 住宅使用双夹墙框架而非 SIP 板建造，这样能够使施工人员能够控制物料配送时间，并调节不可预测现场条件。

The Skinny houses were constructed with double-stud, stickframe construction rather than SIPS. This allowed the field crew to control timing of material delivery and adjust to unpredictable field conditions.

在混凝土与基木板结合处使用垫片

GASKETS WERE USED AT ALL POINTS WHERE CONCRETE MET SILL PLATES

100K住宅具有小巧、高效、超级绿色的特点，是可持续性建筑的典范，是人们可以负担得起的住宅选择。

Small, efficient, and super-green, the 100K Houses provide sustainable, affordable options for homebuyers.

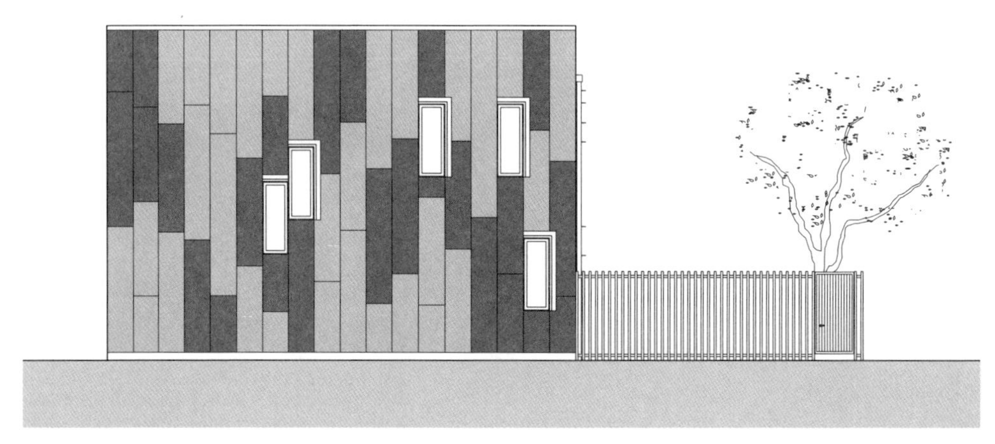

西立面图 WEST ELEVATION

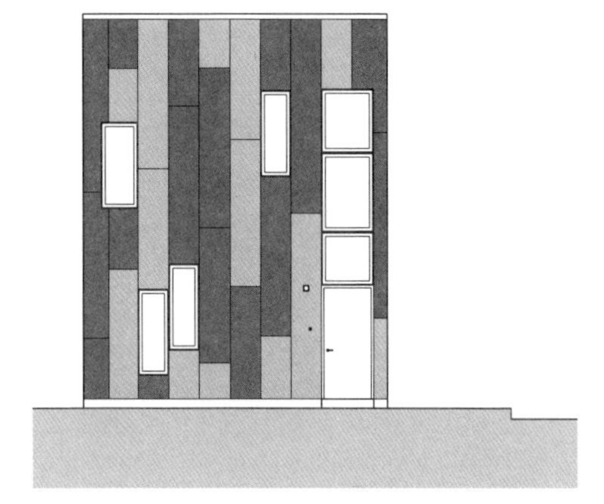

北立面图 NORTH ELEVATION

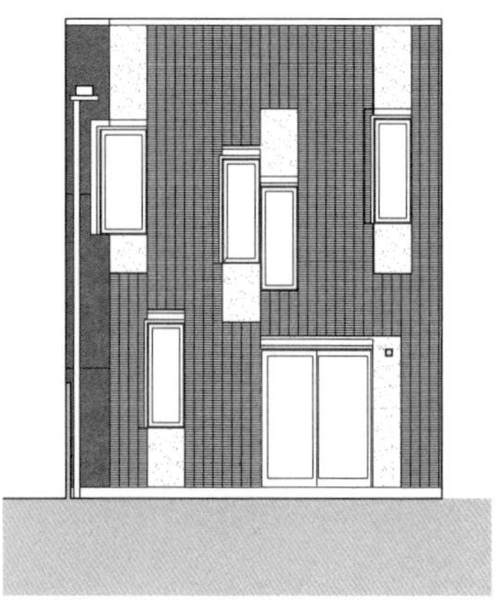

南立面图 SOUTH ELEVATION

2712

海恩堡养老院

NURSING HOME IN HAINBURG

项目名称：

海恩堡养老院

Project Name:

NURSING HOME IN HAINBURG

项目设计： Christian Kronaus + Erhard An-He Kinzelbach
项目地点： 奥地利，海恩堡
立面材料： 纤维水泥板
摄影： Thomas Ott

Architects: Christian Kronaus + Erhard An-He Kinzelbach
Location: Hainburg, Austria
Façade Material: fiber cement
Photographs: Thomas Ott

阳光，沙滩，蓝天，白云，如诗如画的热带风光让无数人心驰神往。椰林摇曳中袅袅走来一位少女，头戴斗笠，身着抹胸短裙，色彩鲜艳亮丽，洋溢着浓郁的热带风情。这样的温暖与惬意，也是老人们晚年所向往的。这座养老院一扫老年建筑的沉闷，建筑表皮用彩色的石棉水泥瓦构成炫丽的像素墙，放眼望去，红黄相间，色彩鲜亮，给老人们平淡的生活增添了一抹色彩，为他们打造一个色彩斑斓的幸福之"家"。

Sunshine, beach, blue sky, white clouds - the picturesque tropical scenery attracts thousands people's thoughts to fly there. A girl, wearing a sun bonnet and a colorful short strapless dress, gently walks through the coconut grove, a strong tropical feeling being permeated. This kind of warmth and comfort is also what the seniors expect for their later years. The Hainburg nursing home gets rid of the dullness of the common buildings for the aged. The façade of the building is covered with colorful Eternit-shingles which form a fascinating pixelated skin. The bright colors bring much vigor to the aged, making this building a colorful, happy home for them.

© 郑亚男（Nancy Zheng）

新建筑垂直安置在原建筑之上，呈双层结构，外立面采用简洁的栅格形式。为了打破单调感，给建筑注入差异和个性的同时，又保持整体的统一性，建筑的表皮就像是变色龙般在粉色和绿色间逐渐变化。东部的新老建筑结合处是粉色，然后逐渐变成绿色，最后西部的新老建筑结合处又变回了粉色。

建筑表皮通过使用两种颜色的菱形石棉水泥瓦构成像素墙。建筑师采用简单的方法，即根据折线的连续反射，增加一种颜色的像素，同时减去另一种颜色的像素。在这个基本原则的指导下，建筑表皮就形成了最终的效果。同时，也使得建筑工人的手工工作和系统安装变得十分简单。

The new building's volume - a double-story, compact bar - is positioned perpendicularly to the existing historic building. In order to break monotony and instill differentiation, while maintaining overall coherence, the skin, like a chameleon, mediates between two predominant colors: pink and green. It gradually changes from one to the other, starting with pink at the eastern end of the joint between old and new, changing to green and changing to pink again when returning to the joint on the western side.

The skin is pixelated through the use of diamond-shaped Eternit-shingles in two colors. A simple algorithm is introduced, based on a combination of successive mirroring at the fold lines and the gradual addition of pixels of one color while subtracting the other. This set of basic rules allows for the described performance. At the same time, it keeps the manual and systematic installation by the construction workers simple.

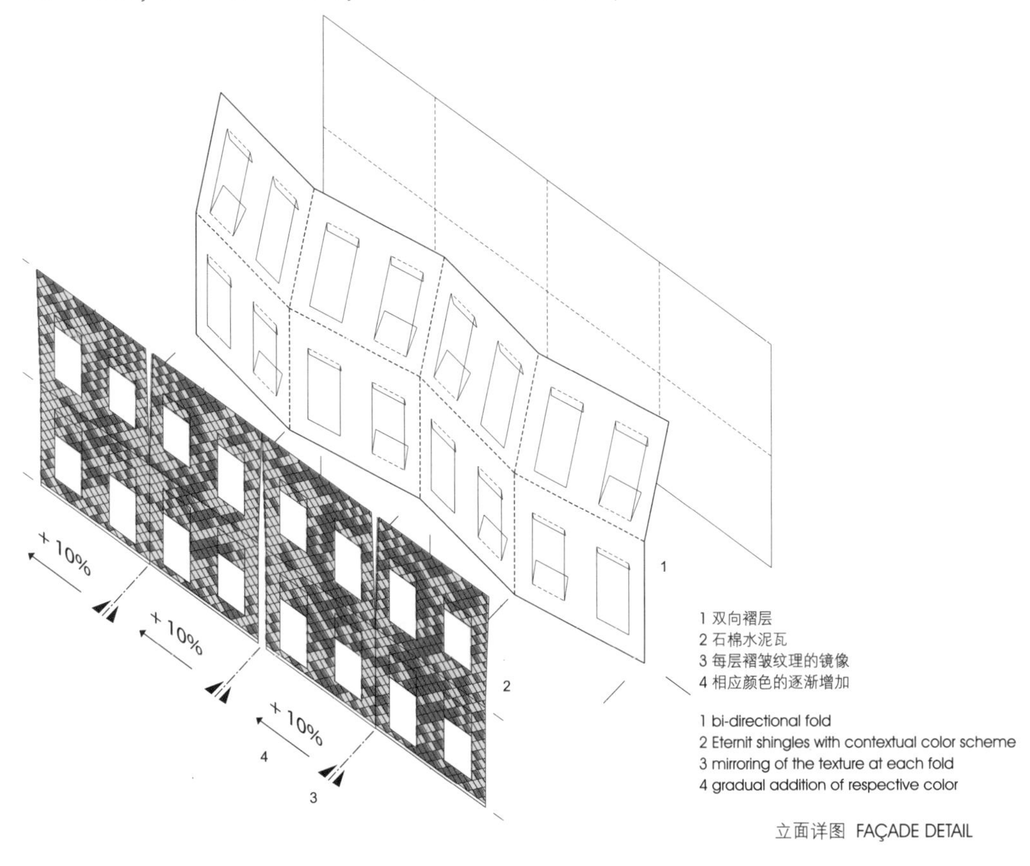

立面详图 FAÇADE DETAIL

随着人均寿命的延长，社会老龄化现象越来越明显，因此老年之家及其建筑结构显得越来越重要。这些建筑将成为老年人的"最后一个家"。老年人在这样的"家"中能够得到多方面的照顾及护理。

In an aging society and due to the tendency in the industrialized west of extended lifetimes, the typology of nursing and retiree homes and its architectural manifestations increasingly gain importance. This happens especially in regards to these buildings' relevance as people's last home in life.

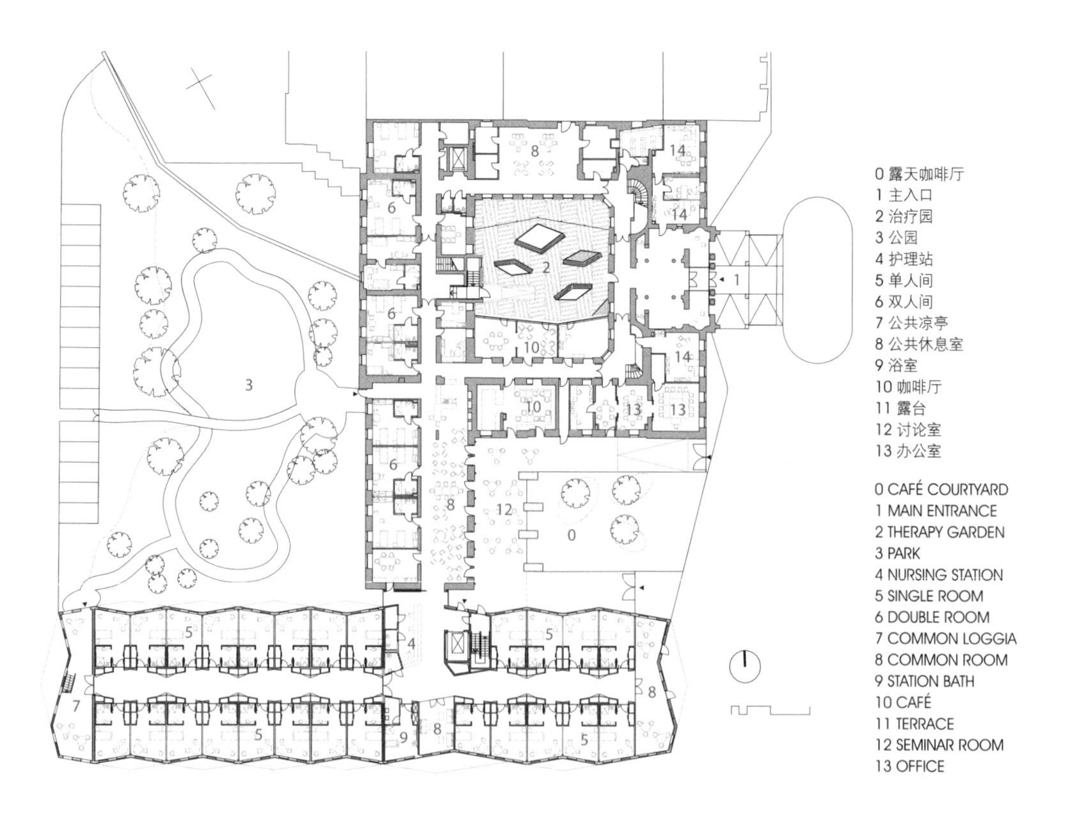

一层平面图 GROUND FLOOR PLAN

艾米迪奥纳瓦罗中学

EMÍDIO NAVARRO SECONDARY SCHOOL

项目名称：

艾米迪奥纳瓦罗中学

Project Name:

EMÍDIO NAVARRO SECONDARY SCHOOL

项目设计： Oficina Ideias em Linha - Arquitectura e Design, Lda.
项目地点： 葡萄牙，阿尔马达
立面材料： 混凝土
竣工时间： 2010 年
摄影： FG+SG - Fernando Guerra, Sergio Guerra, Francisco Nogueira

Architects: Oficina Ideias em Linha - Arquitectura e Design, Lda.
Location: Almada, Portugal
Façade Material: concrete
Completion: 2010
Photographs: FG+SG - Fernando Guerra, Sergio Guerra, Francisco Nogueira

无袖坎肩、紧身短裤搭配金属圆环与镂空黑丝袜，尽展女性的狂野与青春。红与黄的激情碰撞，如燃烧的日晖，热情奔放。腰间如闪电般的饰带，更起到了画龙点睛的作用，成为视觉上的一大亮点。这座中学体育馆的外立面由立柱呈"V"字形搭接而成，律动感十足，犹如保护壳一样包裹着内侧建筑。内部灯光明亮，展现了学生们青春向上的生命力，朝气蓬勃。

Sleeveless vest, hot pants, metal rings and black hallow-out stockings fully present the vigor and wildness of female. The combination of the red and yellow colors is as enthusiastic as the burning sunshine. The lightening-like belt around the waist adds the finishing touch and becomes a visual highlight. Several V-shape columns compose the outer wall of the school sports center, protecting the inner space like a shell, and creating a sense of rhythm and movement. The bright light glows from inside, reflecting the vitality and vigor of the students.

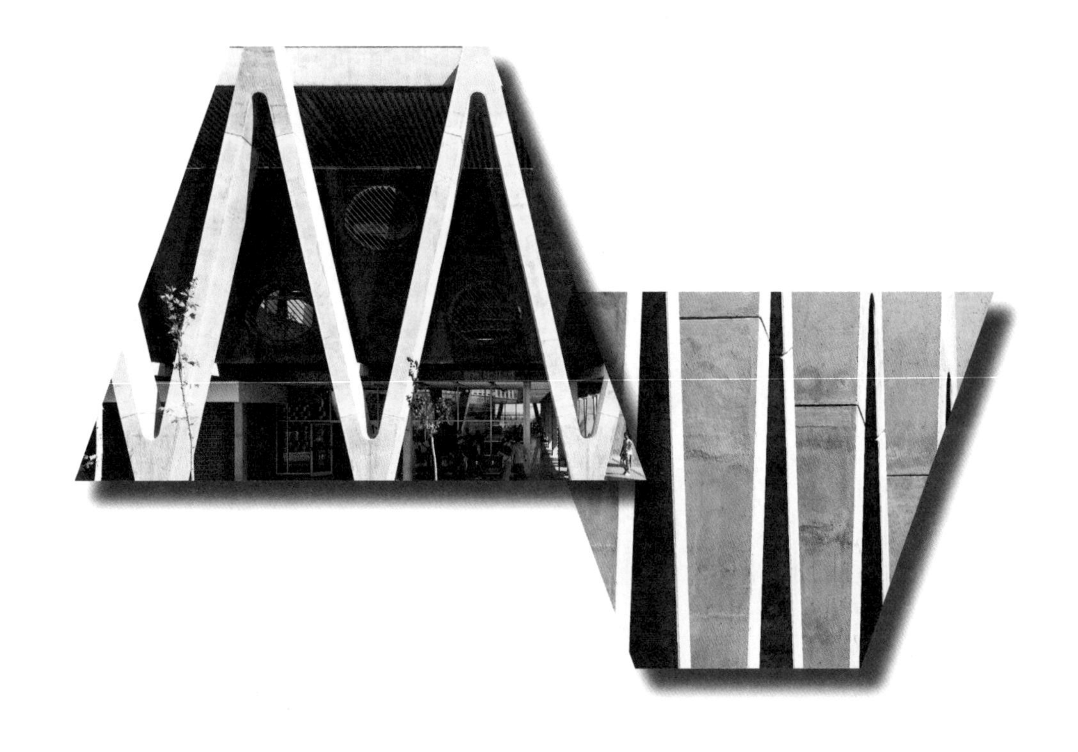

© 宋宁（Bella Song）

"面料定制"
"Material"

由于受场地面积所限，因此体育馆使用了上宽下窄的设计形式，保证了上方的结构平衡。预制混凝土结构形成大型外壳，通过利用一系列有规则排列的倾斜支柱来分散支撑力。

The sports pavilion layout was designed according to the needs of a large structural balance on the upper level and the construction of the tightened areas below. The precast concrete structure was designed as a large scale exoskeleton, distributing charges and supports through a regular array of inclined pillars.

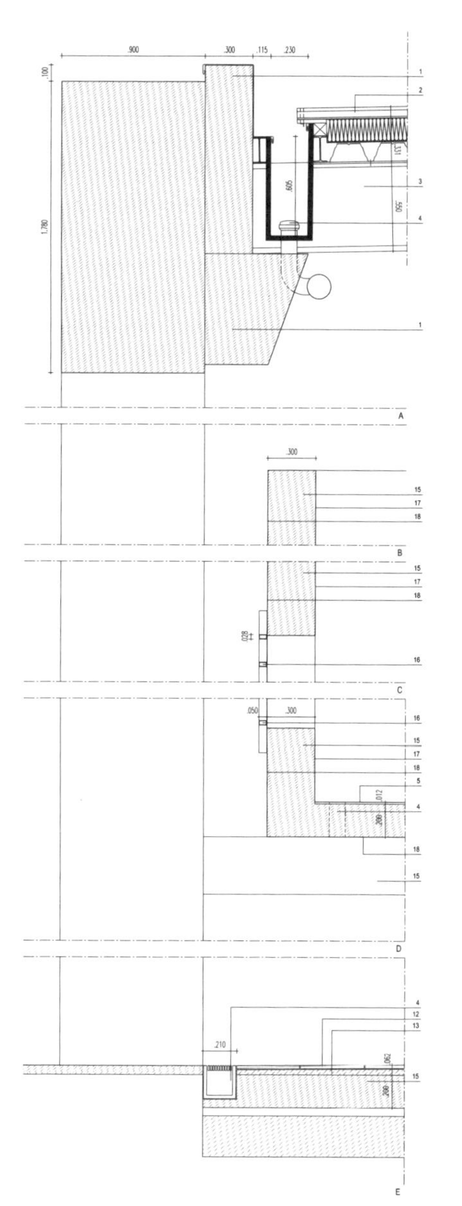

1 预置混凝土结构
2 屋顶 Kalzip 系统
3 金属结构
4 排水系统
5 运动场地板
6 石棉隔热板 80 mm
7 铝合金窗户
8 穿孔清声石膏板顶棚
9 石膏板顶棚
10 Terrazzo 瓷砖
11 水泥板
12 粉刷石膏
13 外部沥青铺面
14 通风系统
15 混凝土结构
16 金属支架
17 白色混凝土结构
18 黑色混凝土结构

1 Precast Concrete Structure
2 Roof "Kalzip" System
3 Metallic Structure
4 Drainage
5 Sports Floor
6 "Rockwool" Thermal Insulation 80 mm
7 Aluminum Windows
8 Perforated Acoustic Gypsum Board Ceiling
9 Gypsum Board Ceiling
10 "Terrazzo" Tiles
11 Cement Screed
12 Painted Plaster
13 Exterior Bituminous Paving
14 Ventilation System
15 Concrete Structure
16 Metallic Grill
17 Concrete Structure Painted White
18 Concrete Structure Painted Black

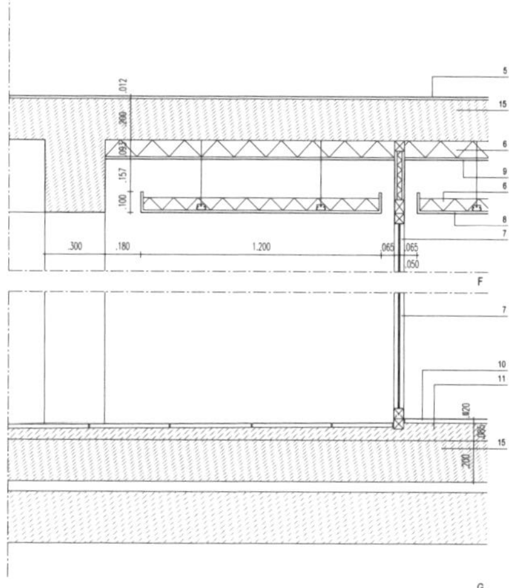

该结构与另一混凝土结构相辅相成，排列规则的柱/梁体系构成的结构支撑着上层。此外，混凝土"盒"状结构包裹着室内运动空间。建筑顶部由用钢梁和穿孔结构钢板构成系统的外层结构所支撑，表面覆有复合铝板，可以起到隔热和隔声作用。

This structure is reinforced by a second one - also in concrete - characterized by a regular grid of pillar / beam system which supports the upper level and the concrete wall box that envelopes the indoor sports area. The roofing is supported by the outer structure laying over a steel beam and perforated steel structural sheets system and finished with aluminum composite panels considering thermal and acoustic insulation.

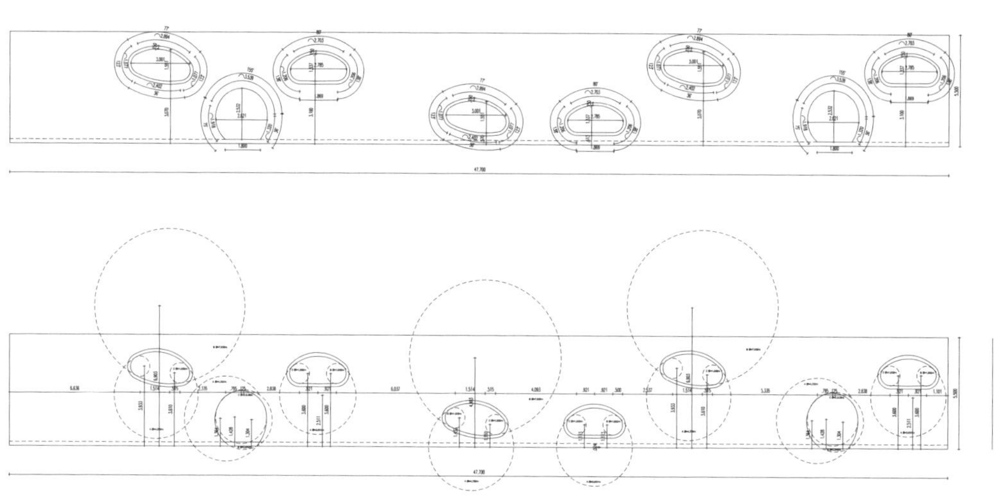

体育馆立面图 PAVILION ELEVATIONS

西立面图 WEST ELEVATION

由于学校周边条件的限制，这座室内体育馆占地面积仅为24.00 m x 44.00 m，因此建筑师采取特殊的地形策略来容纳运动区域。把体育馆的地面降低，使运动区与周围建筑处于同一水平，此外下层还设有厨房、餐厅、宴会厅等。人们可通过两座混凝土天桥到达运动区。

Given the constraints of the existent school perimeter on housing the indoors sports pavilion (24.00 m x 44.00 m), a topographical strategy was adopted in order to accommodate the sports area through lowering grounds and therefore allowing the leveling of the new platform with the surrounding buildings complemented by a new platform on lower grounds where the facilities such as kitchen, refectory and convivial areas are located. The accessibility to the sports platform is done through two concrete bridges.

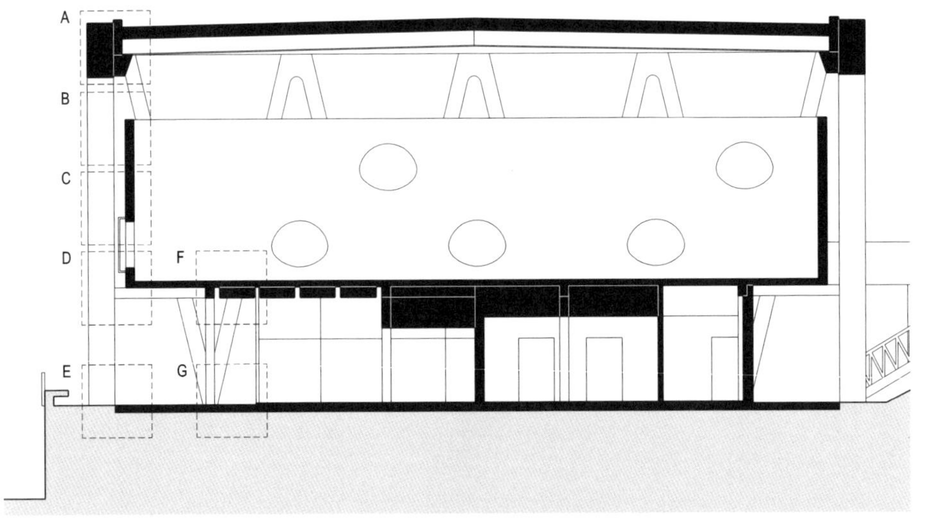

剖面图 SECTION

科伯格别墅

COBOGÓ HOUSE

项目名称：

科伯格别墅

Project Name:

COBOGÓ HOUSE

项目设计： Studio mk27
项目地点： 巴西，圣保罗
立面材料： 混凝土
竣工时间： 2011 年
摄影： Nelson Kon

Architects: Studio mk27
Location: Sao Paulo, Brazil
Façade Material: concrete
Completion: 2011
Photographs: Nelson Kon

几何图案，是设计中不可或缺的要素，对服装设计、建筑设计等领域影响甚巨。以几何图案点缀于服装，再简单的款式都不会显得单调。立体化的几何图案更容易引起视觉上的冲击，特别是“规则设计”的几何图案运用于服装中，规律中加重了变化的层次感，时尚而设计感十足。同样，以几何图案装饰建筑立面，亦有异曲同工之妙。科伯格别墅的外墙以抽象简约的几何图案形成“空间化”的镂空孔洞，连绵不断，深深吸引着众人的眼球。

Geometric pattern is an essential element in design. It has a great influence on fashion design, architecture design and so forth. Adorned by geometric pattern, the clothes won't be monotonous no matter how simple the style is. 3D geometric pattern is much easier to arouse visual stimulation. In particular, when "regularly designed" pattern is applied on clothes, regularity enhances the varied layers and makes the clothes appear fashionable. It's the same when designing a building's façade with geometric pattern. The external wall of Cobogó House is comprised of spatialized hollow-out elements which are formed by abstract, simple geometric patterns. Its infinity attracts eyeball of all people.

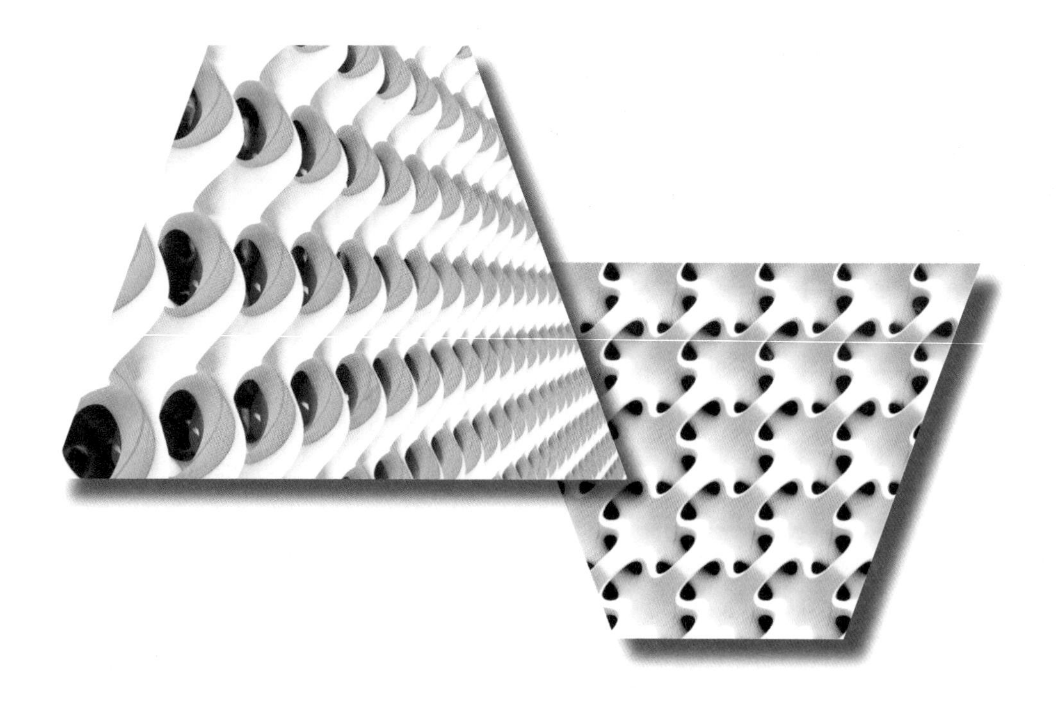

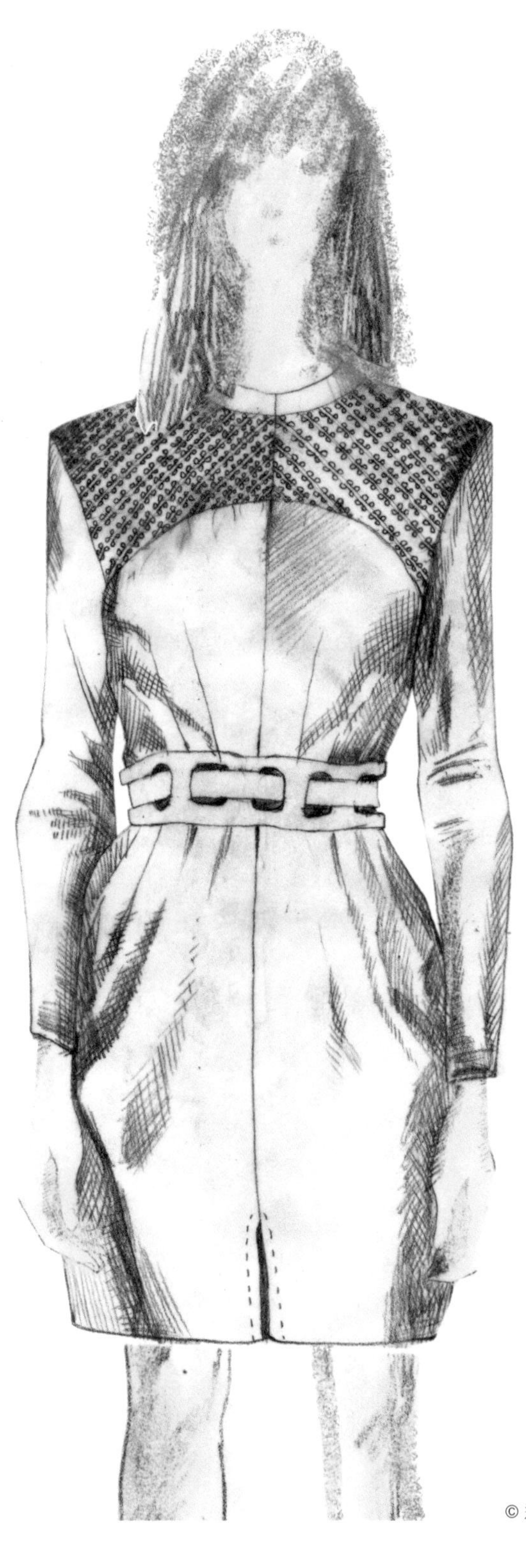

© 郑亚男（Nancy Zheng）

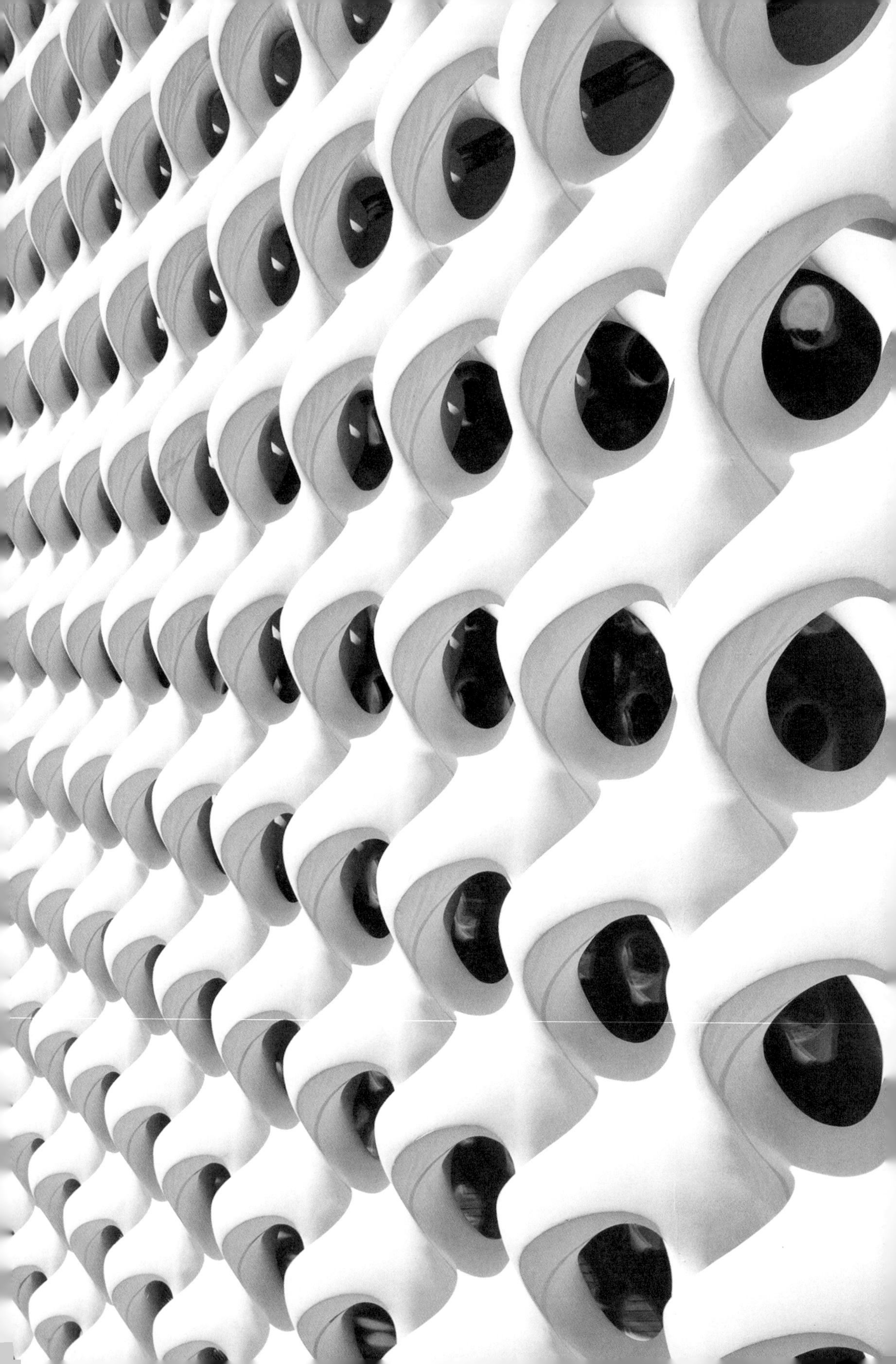

大量的热带阳光洒在这所房子顶部的“白色方盒子”上，透过上面的镂空把斑驳的光影洒落在室内的地板上。墙面上这些“空间化”的镂空孔洞通过光影的勾勒变得更加立体，加上环境的影响使其效果加倍明显且多变。每一天，每个月，这些孔洞元素会根据太阳光线入射角度的不同而呈现出不同形态。而在夜晚，其效果再次变幻，灿烂的灯光瞬间打破夜幕，让建筑变成一个熠熠生辉的“珠宝盒”。

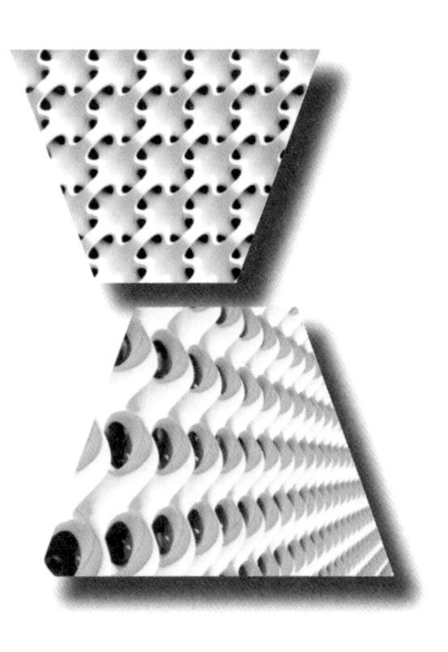

The light of the abundant tropical Sun falls on the white volume of the top floor of the house, penetrating the holes of the hollowed elements and covering the floor of the interior space. Thus, the design of spatialized lace is formed from the shadows and solar rays. The effect is multiplied throughout the ambient, making a construction from the light itself. Throughout the days, throughout the months, the hollowed-out elements take on different forms with the incidence of the sun. At night, this effect once again is transformed, the light spill through the façade from the interior, make it a shining treasure box in the darkness.

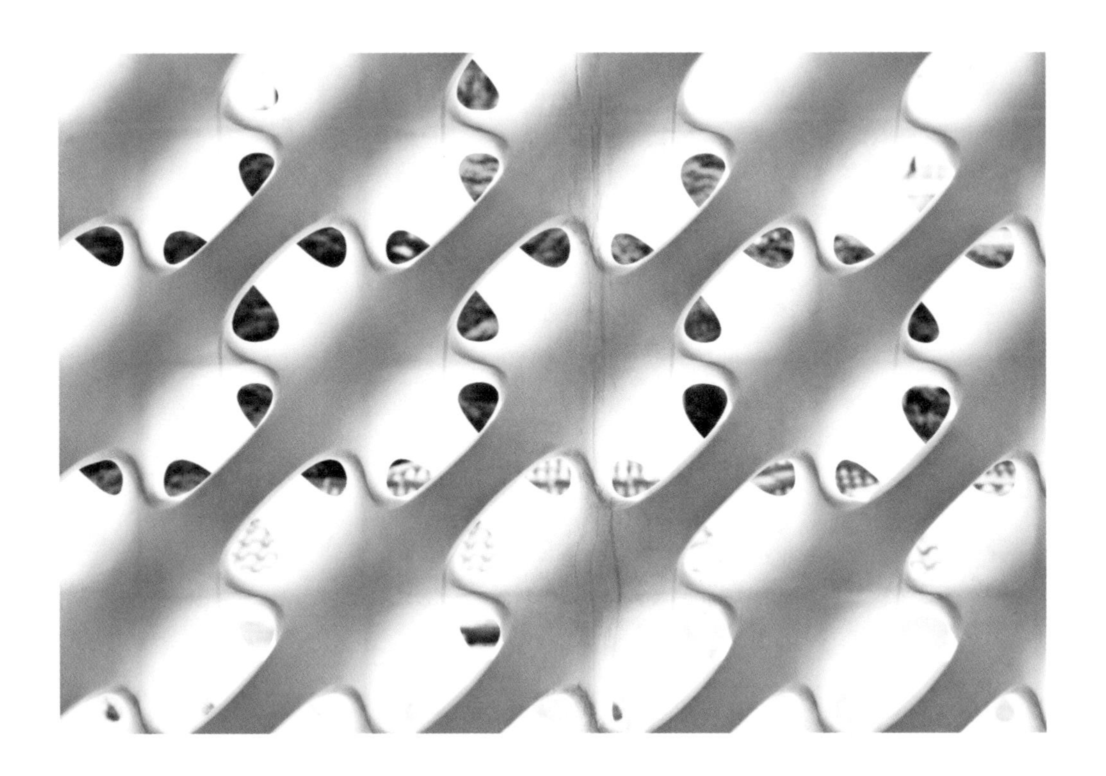

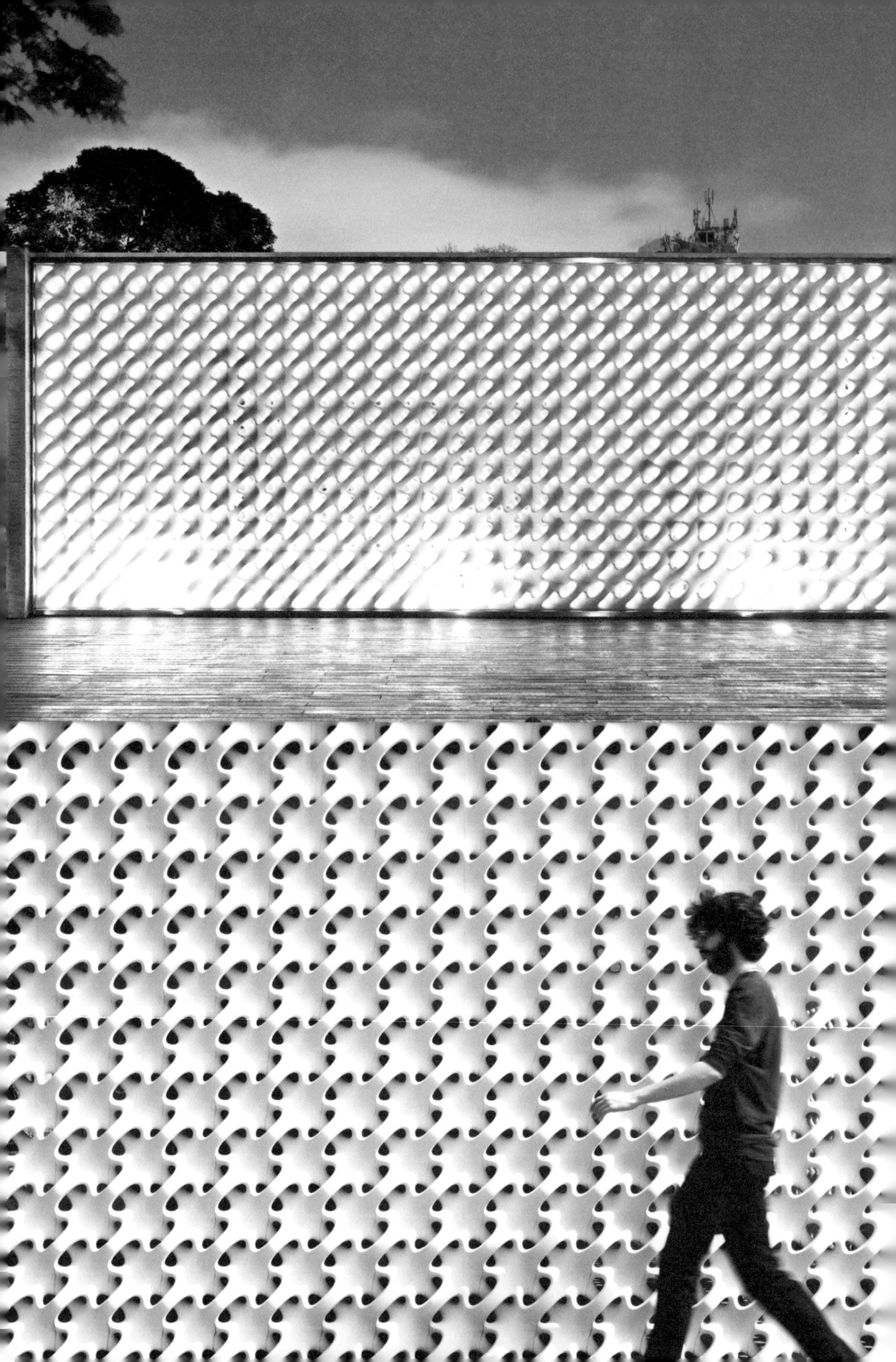

建筑的外墙由镂空元素构成，其几何形状由无限的曲线交织而成，形成复杂的结构。这种模块元素是由欧文·豪尔创作的艺术作品。豪尔自 1950 年以来便为建筑空间创作一些几何雕塑。他创作的抽象简约的几何图案与建筑形成对话，并让人们很容易联想到巴西现代建筑的踪迹。弯曲的线条，借鉴了尼迈耶的巴西利亚建筑。另外，混凝土模块源自 Cobogó，由路西奥·科斯塔借鉴殖民时期的建筑，并且将其推广开来。而 Cobogó 也成为这座别墅名字的由来。

The soft volumetric geometry of the hollowed-out elements comprising the walls is a complex construction, made with infinite curved lines. The modular element, a work of art, was designed by Erwin Hauer who, since 1950, has conceived and made sculptures for architectural space. His minimalist elements dialogue with the architecture and remind us of some traces of Brazilian modern architecture. The curved lines, designed with perfection nod to the architecture of Brasilia by Niemeyer, furthermore, the concrete modules descend from the Cobogós - which lends its name to the house - created in Recife and diffused by Lucio Costa in delicate references to colonial architecture.

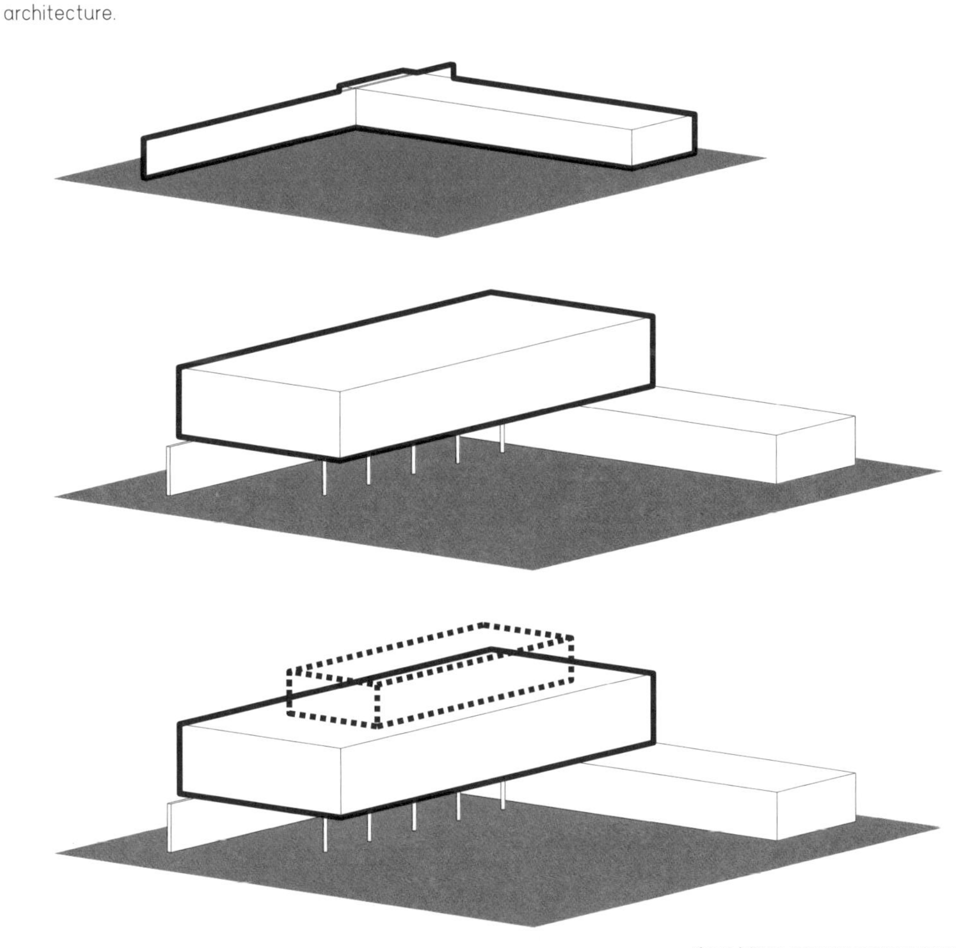

施工过程图 CONSTRUCTION PROCESS

科伯格别墅是一座现代住宅，很自然地将欧文·豪尔的艺术融入其中，作为建筑的一部分。上方的体量由腻子、混凝土、木材构成，与露台花园一同，由欧文·豪尔设计的镂空元素构筑。

The Cobogó House is a modern house in which the art of Erwin Hauer can be used naturally, as a part of the entire architecture. In the play of mounting pure volumes, made of white spackling paste, concrete and wood, lies, together with the terrace garden, the volume constructed from the hollowed elements by Erwin Hauer.

1 客厅	1 Living Room
2 餐厅	2 Dining Room
3 厨房	3 Kitchen
4 电视房	4 TV Room
5 烤肉区	5 Barbecue
6 阳台	6 Terrace

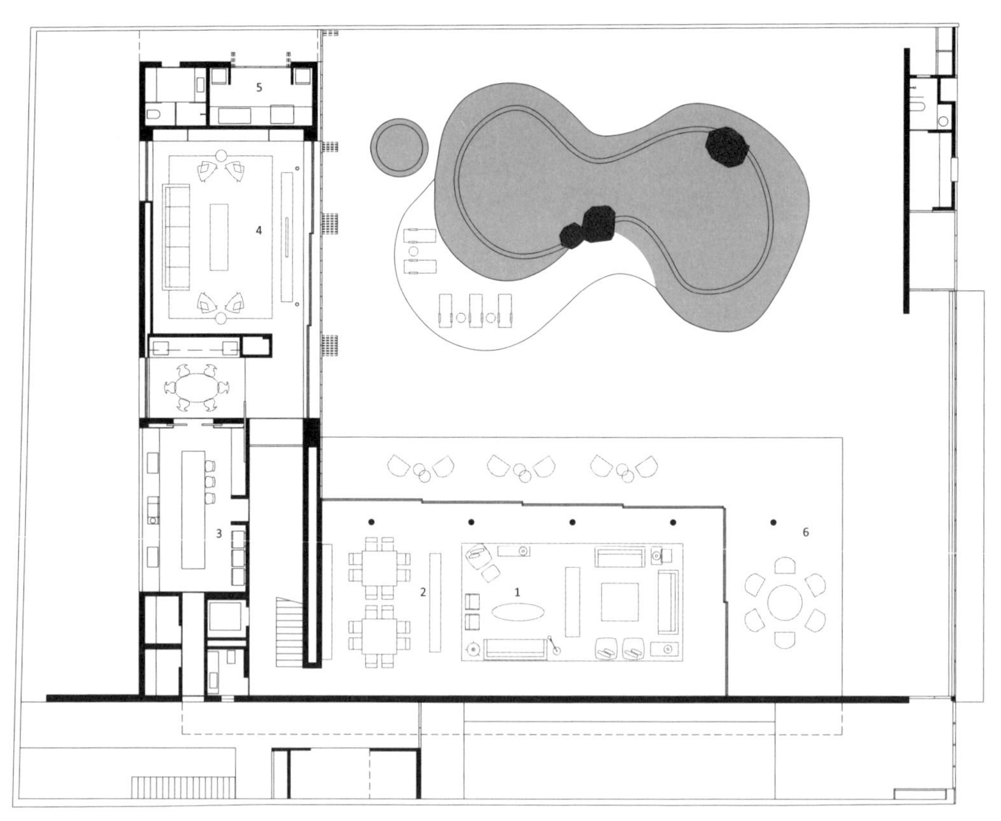

一层平面图 GROUND FLOOR PLAN

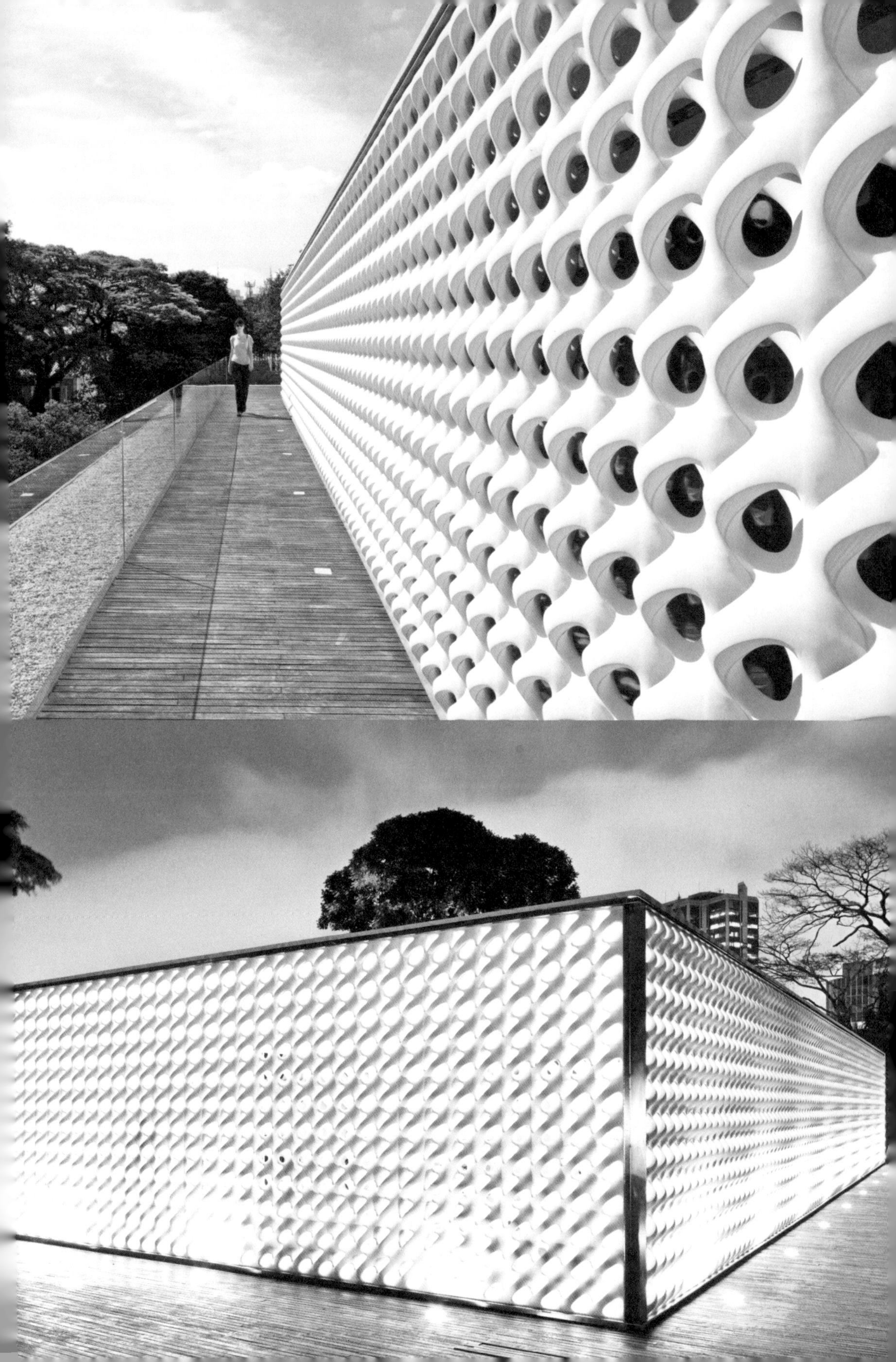

皮埃尔·布丁街托儿所

CRÈCHE RUE PIERRE BUDIN

项目名称：

皮埃尔·布丁街托儿所

Project Name:

CRÈCHE RUE PIERRE BUDIN

项目设计： Emmanuel Combarel Dominique Marrec architectes (ECDM)
项目地点： 法国，巴黎
立面材料： 混凝土
竣工时间： 2012 年
摄影： Luc Boegly, Benoît Fougeirol

Architects: Emmanuel Combarel Dominique Marrec architectes (ECDM)
Location: Paris, France
Façade Material: concrete
Completion: 2012
Photographs: Luc Boegly, Benoît Fougeirol

白色，象征着纯洁与美好。童话中的白雪公主是每个女孩的梦，那一袭纯净的白裙让每个女孩为之向往。优雅的长裙完美地勾勒出曼妙的身段，纤腰盈盈尽显女人风情，妖娆中又似清纯无比。随风起舞，裙摆飞扬，仿佛是花中的精灵，又如纯洁的天使。这座幼儿建筑通体洁白，立面呈凹凸起伏的曲面，形成波浪起伏的韵律感。

White represents pure and fine. Being a princess like Snow White in the fairy tale is every girl's dream, and a pure white dress makes every girl yarn for. The graceful dress perfectly articulates her stunning figure, and the slim waist fully reveals the girl's charm, coquettish, but also pure. Dancing in the wind, with the hemlines swirling around the ankle, she likes a Flower fairy, or an angel. This kindergarten is painted white all over. Its rippling façade creates an undulating rhythm.

© 郑亚男（Nancy Zheng）

该建筑是一个托儿所项目。建筑的外墙由预制混凝土面板构成，形成一种波浪起伏的效果，宛如水面上的涟漪一般。外墙只有一侧装有透明的彩色窗户，其高度不一，能够满足不同人群的视野要求。在这里，无论是大人还是小孩，无论是父母还是工作人员，都能感到温馨与关怀。

This white building is a nursery school. The wall of the building is made of prefabricated concrete panels, creating an undulating effect. It's like ripples on the surface of water. The surrounding wall is drilled by translucent and colored windows. These windows have various heights, for a place thought as much for the children than for the adults, the parents or the staff.

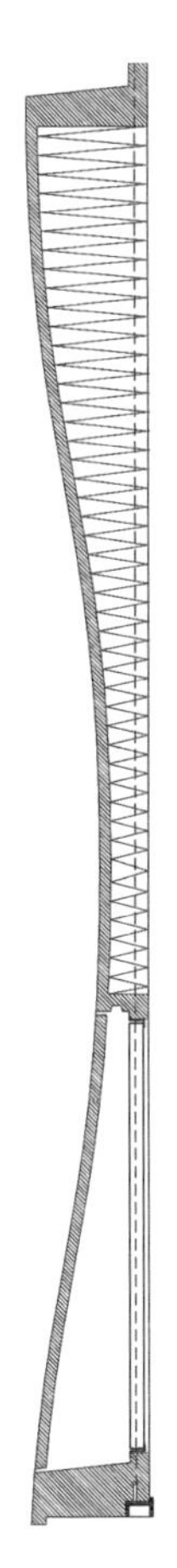

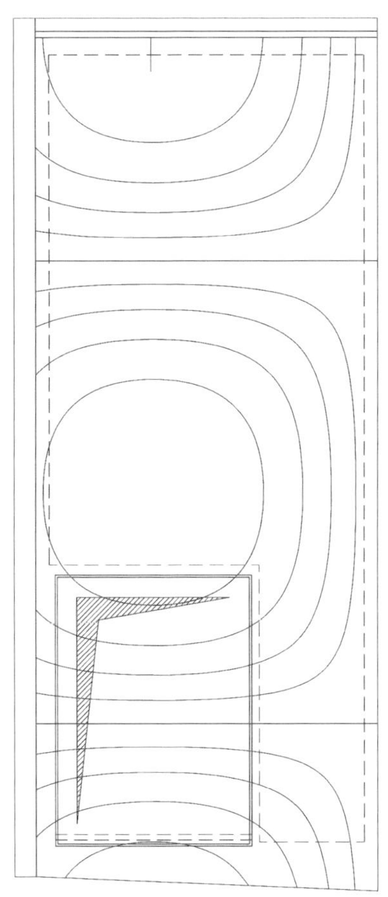

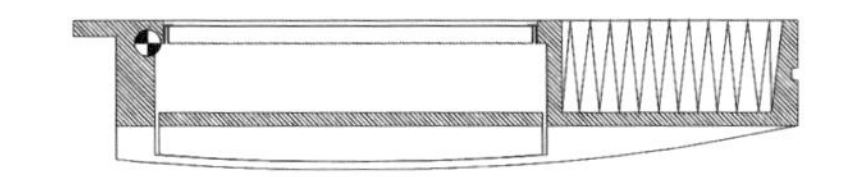

面板详图 PANEL DETAIL

这座托儿所建筑规模较小。其规模源自项目对面积和体积的要求，而建筑内涵源自于其特色。这座建筑具有防护性、内敛性的特点，与室外空间相互呼应。建筑共两层，通过精心设计以最大限度获取光线和日照，避开了周边建筑投下的阴影。

The program of the day-nursery introduces a small size, a small scale. If the volume comes from the requirements of the project concerning surfaces and scale, the writing of the building results from its specificity. Protective and introverted, it occupies the ground, interacts with the outside spaces. Developed on two levels, it is organized to get the maximum of light and sunshine, and to by-pass the shade of the giant nearby building.

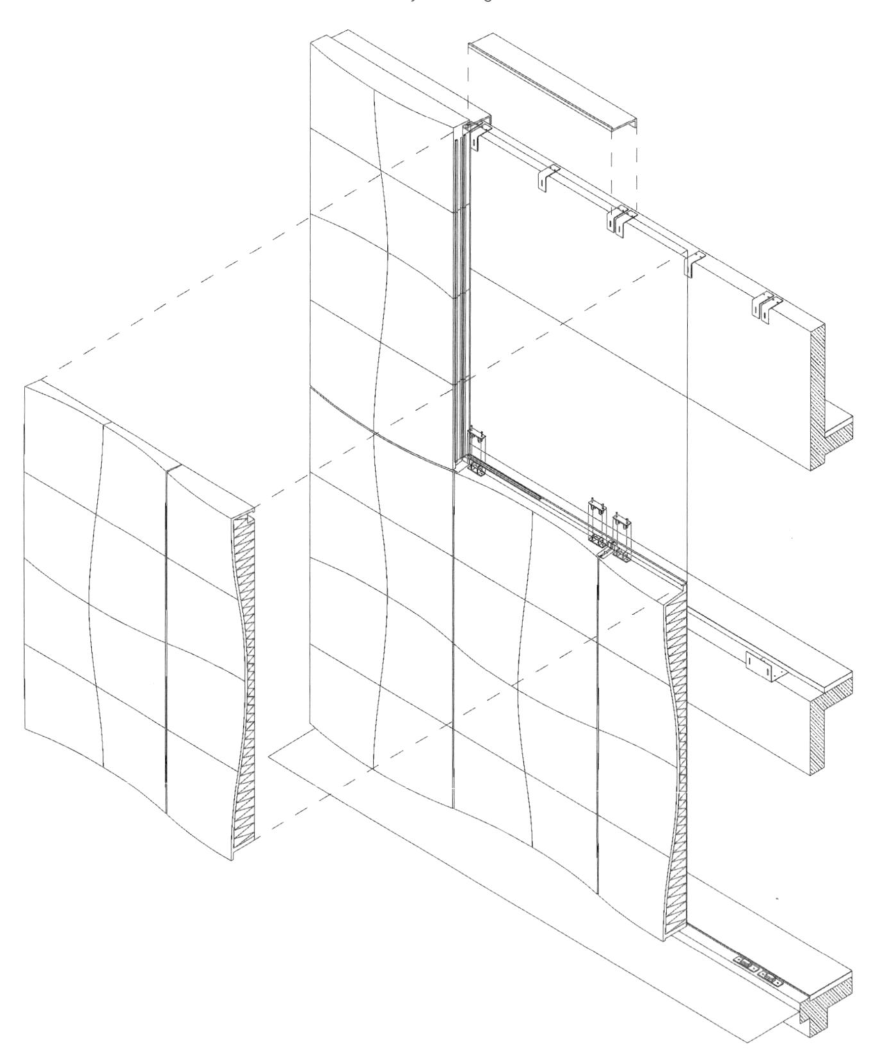

立面详图 FAÇADE DETAIL

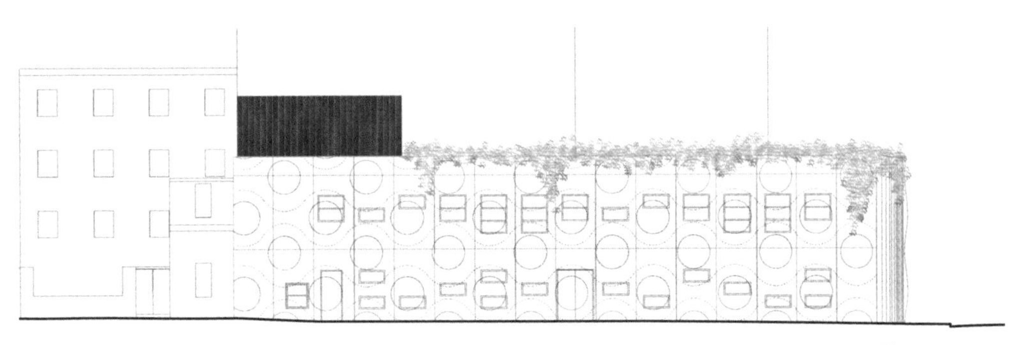

立面图 ELEVATION

该项目位于巴黎近郊区。这是一个多元化地区，建筑的体量、风格、时期各不相同，呈现出古朴、混合、凌乱的氛围，具有典型的巴黎近郊异构结构建筑风格。该建筑以其现代性与当前的杂乱无章相呼应。建筑呈非常规尺寸，打破边界限制（增添都市感），避免流于普通、毫无特色。托儿所建筑处于一座 12 层建筑的挤压和阴影中，这座高大建筑遮挡了光线，并且向外悬伸。因此，这座托儿所是在这种不利条件下建造起来的公共建筑，这是一种新的尝试。

The project takes place into a heterogeneous district made of buildings of any sizes, of any styles, any periods. It's an environment slightly old-fashioned, hybrid and disintegrated, typical of the heterogeneous architecture which characterizes the Parisian peri-urban zones. Modernity came to complete this disorder. Adjacent to the site, an out of size construction, built in derogation of the property limits (adding a supplementary urban intention parameter), forbids any common denominator, any possibility of creating a homogeneous composition. The day-nursery is thus an attempt, for a tiny building of public utility, to exist in an unfavorable relationship in the shade of a twelve story construction which takes light, overhangs and crushes everything.

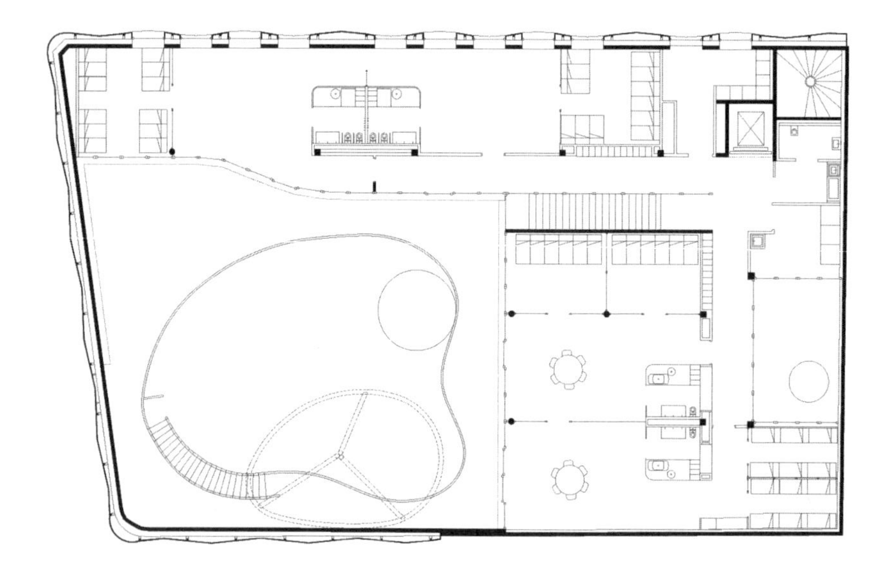

平面图 PLAN

SAUF

图书在版编目（CIP）数据

建筑“时装”定制．混凝土：汉英对照 / 凤凰空间·北京编．-- 南京：江苏科学技术出版社，2013.5
ISBN 978-7-5537-1026-6

Ⅰ．①建… Ⅱ．①凤… Ⅲ．①装饰墙－建筑装饰－混凝土结构－立面造型－世界－图集 Ⅳ．①TU227-64

中国版本图书馆 CIP 数据核字（2013）第 064349 号

建筑“时装”定制——混凝土

编　　者　凤凰空间·北京
责任编辑　刘屹立
特约编辑　田　静　李　瑶
责任校对　郝慧华
责任监制　刘　钧

出版发行　凤凰出版传媒股份有限公司
　　　　　江苏科学技术出版社
出版社地址　南京市湖南路1号A楼，邮编：210009
出版社网址　http://www.pspress.cn
经　　销　凤凰出版传媒股份有限公司
印　　刷　北京建宏印刷有限公司

开　　本　710 mm×1 000 mm 1/16
印　　张　12
字　　数　96 000
版　　次　2013年5月第1版
印　　次　2013年5月第1次印刷

标准书号　ISBN 978-7-5537-1026-6
定　　价　58.00元